Studies in Medieval History and Culture

Edited by
Francis G. Gentry
Professor of German
Pennsylvania State University

A Routledge Series

Studies in Medieval History and Culture

Francis G. Gentry, *General Editor*

Pestilence in Medieval and Early Modern English Literature
Bryon Lee Grigsby

Race and Ethnicity in Anglo-Saxon Literature
Stephen J. Harris

Aspects of Love in John Gower's Confessio Amantis
Ellen Shaw Bakalian

The Medieval Tradition of Thebes
History and Narrative in the OF Roman de Thèbes, Boccaccio, Chaucer, and Lydgate
Dominique Battles

Worlds Made Flesh
Reading Medieval Manuscript Culture
Lauryn S. Mayer

Empowering Collaborations
Writing Partnerships between Religious Women and Scribes in the Middle Ages
Kimberly M. Benedict

The Water Supply System of Siena, Italy
The Medieval Roots of the Modern Networked City
Michael P. Kucher

The Epistemology of the Monstrous in the Middle Ages
Lisa Verner

Desiring Truth
The Process of Judgment in Fourteenth-Century Art and Literature
Jeremy Lowe

The Preaching Fox
Festive Subversion in the Plays of the Wakefield Master
Warren Edminster

Non-Native Sources for the Scandinavian Kings' Sagas
Paul A. White

Kingship, Conquest, and Patria
Literary and Cultural Identities in Medieval French and Welsh Arthurian Romance
Kristen Lee Over

Saracens and the Making of English Identity
The Auchinleck Manuscript
Siobhain Bly Calkin

Traveling through Text
Message and Method in Late Medieval Pilgrimage Accounts
Elka Weber

Between Courtly Literature and Al-Andalus
Matière d'Orient and the Importance of Spain in the Romances of the Twelfth-Century Writer Chrétien de Troyes
Michelle Reichert

Maps and Monsters in Medieval England
Asa Simon Mittman

Rooted in the Earth, Rooted in the Sky
Hildegard of Bingen and Premodern Medicine
Victoria Sweet

"She, This in Blak"
Vision, Truth, and Will in Geoffrey Chaucer's Troilus and Criseyde
T. E. Hill

Through the Daemon's Gate
Kepler's Somnium, *Medieval Dream Narratives, and the Polysemy of Allegorical Motifs*
Dean Swinford

Through the Daemon's Gate

Kepler's *Somnium*, Medieval Dream Narratives, and the Polysemy of Allegorical Motifs

Dean Swinford

Routledge
New York & London

Transferred to digital printing 2010

Routledge
Taylor & Francis Group
270 Madison Avenue
New York, NY 10016

Routledge
Taylor & Francis Group
2 Park Square
Milton Park, Abingdon
Oxon OX14 4RN

International Standard Book Number-10: 0-415-97764-9 (Hard), 0-415-88607-4 (Soft)
International Standard Book Number-13: 978-0-415-97764-7 (Hard), 978-0-415-88607-9 (Soft)

Visit the Taylor & Francis Web site at
http://www.taylorandfrancis.com

and the Routledge Web site at
http://www.routledge-ny.com

For Bill and Phyllis Swinford

Contents

Series Editor's Foreword

As the name suggests, *Studies in Medieval History and Culture* is a series of studies in all disciplines dealing with the broadly-conceived, primarily European, medieval era. It is comprised of outstanding recent dissertations as well as monograph-length studies. The editor also welcomes suggestions for essay volumes focusing on an interdisciplinary theme. Thus it is open to scholars at every stage of their careers.

The interest in Medieval Studies, which underwent a dramatically broad renascence in the 1960s and 1970s, shows, even now at the beginning of the 21st century, no sign of abating. Scholars, both junior and senior, are consistently producing works of original research of the highest caliber. Indeed, Medieval Studies enters the new century as fresh and vigorous as never before. Just the merest of glances at the volumes already published in *Studies in Medieval History and Culture* offers evidence to support this claim. Previously published studies include analyses of individual works and authors of Latin and vernacular literatures, historical personalities and events, theological and philosophical issues, as well as new critical and theoretical approaches to medieval literature and culture.

Great changes in teaching and research have occurred in the broad area of Medieval Studies in the past several decades and continue to do so. *Studies in Medieval History and Culture* seeks to facilitate intellectual exchange in the field by providing an outlet for scholars to disseminate the results of their research, while at the same time pointing to and highlighting those new directions that will shape and define ongoing scholarly discourse.

Acknowledgments

I wish to thank, first and foremost, Jim Paxson, who went above and beyond the call of duty as an advisor and friend. Thanks, Jim, for everything. I also want to thank Al Shoaf, Nora Alter, Phil Wegner, and Bob Hatch for their insight and advice on earlier drafts of this book. I'd also like to extend thanks to the Fulbright Commission and Maggie Nicholson, the director of the Fulbright program in Belgium, for the tremendous opportunity to participate in this extraordinary program. I wrote the initial draft of this book while on a Fulbright Fellowship in Belgium and could not have formulated many of these ideas without the experiences of that year. I'd also like to thank Fernand Hallyn for his help, warmth, and hospitality in Belgium. You are an inspiration to me. Benoit de Baere, Alexander Roose, and Caroline de Mulder all deserve gratitude for helping me to feel at home in Gent. To my friend Jason Flom, the many hours of hiking and conversation we have shared in multiple countries help me to order my mind and soothe my soul. Finally, thanks go to Alison Van Nyhuis, the sun at the center of my universe.

Chapter One

Polysemy and Allegorical Signification

MANDEVILLE ON COSMOGRAPHY

Sir John Mandeville, that travel writer esteemed for his reliability by a medieval and early modern audience that included Leonardo da Vinci and Christopher Columbus, has the following to say about the shape of the earth and its position in the universe: "For, as I said before, God made the earth quite round, in the middle of the firmament."[1] Though Mandeville attests to the finitude of the earth and calculates its circumference, he informs us that he "cannot speak properly" of Earthly Paradise "for I have not been there and that I regret."[2] Still, he confirms as valid the reports from "trustworthy authorities" that the Earthly Paradise "is so high it touches the sphere of the moon."[3] With this statement, Mandeville evokes Paradise as the point where the earth comes into contact with an outside universe composed of interlocking spheres.

Building on an established tradition, Mandeville does not, of course, invent the geocentric model of the universe. However, I begin with this point in *The Travels of Sir John Mandeville* because of Mandeville's comments regarding the modes of transport available to the traveler desiring to vacation in Paradise. Not content to tell us that "no living man can go to Paradise,"[4] Mandeville instead points out the inadequacy of travel by land and sea:

> By land no man can go thither because of the wild beasts in the wilderness, and because of the hills and rocks, which no one can cross; and also because of the many dark places that are there. No one can go there by water either, for those rivers flow with so strong a current, with such a rush and such waves that no boat can sail against them.[5]

The only acceptable mode of transportation is supernatural: "And so no man, as I said, can get there except through the special grace of God."[6] Terrestrial vehicles cannot provide access to either the Earthly Paradise or the encircling celestial bodies. The special grace of God, however, supplied by faith, allows the elect to soar beyond natural constraints. While Mandeville maintains a firmly terrestrial focus in his travel guide, medieval writers such as Alain de Lille, Bernardus Silvestris, and Hugh of St. Victor, seeking to reconcile the works of newly translated classical authorities with Christian philosophy, composed cosmological allegories that provided detailed accounts of flights beyond the earthly sphere.

Between the initial appearance of Mandeville's travels in 1356 and Copernicus' *De Revolutionibus Orbium Caelestium* in 1543, however, scientific developments necessitated a reinterpretation of the position of the earth in the universe. Still, the acceptance of Copernicanism took time, so that, even as astronomers such as Galileo Galilei and Johannes Kepler accepted the idea of a heliocentric universe, social institutions, including the church and universities, transmitted and inculcated views clearly based on Neoplatonist models. Improved observations based on the telescope helped to invalidate the geocentric universe and also transformed the imaginative possibilities of travel. Thus, Young's exuberant proclamation in *Night Thoughts* that

> The soul of man was made to walk the skies,
> Delightful outlet of her prison here!
> There, disencumber'd from her chains, the ties
> Of toys terrestrial, she can rove at large,
> There, freely can respire, dilate, extend,
> In full proportion let loose all her powers;
> And, undeluded, grasp at something great[7]

differs markedly from Mandeville's own conception of access to empyreal realms, and this despite Mandeville's preoccupations with the fantastic. Marjorie Nicolson cites the development of the telescope as the main factor precipitating this imaginative shift. But how does a scientific device replace divine grace as an appropriate mode of transport to the stars? Or, more importantly, how is this shift represented in literary and scientific texts?

Michel Serres poses a similar question in *La Légende des Anges,* wondering, "by what partial understandings do the natural sciences divest themselves of the enchantment of the supernatural?"[8] Serres' phrase, "partial understandings," emphasizes the slow process of the movement from supernatural to scientific. However, while many historians of science may see this slow process as one that culminates in the seventeenth century, I argue, influenced by Serres, for the continuous, and continuously incomplete, nature of this process.

MAPPING ALLEGORY: AN OVERVIEW

The vehicles of supernatural voyages, in keeping with Michel de Certeau's use of the term, may rightly be called *metaphorai.* Such a vehicle, emblematic of a moving gateway, its initial dimensions corresponding to the measurements of the celestial spheres, enacts the passage from supernatural to natural. While *metaphorai* can, perhaps, describe any vehicle or boundary, the narratives of early modern science employ vehicles as narratives in pursuit of an ontological repositioning of the natural world. My interdisciplinary exploration of literature and science, then, seeks to understand how phenomena are described in metaphoric terms. I propose a semiotic analysis of the narrative markers used to describe not just a new world, but also a new way of describing new worlds.

This study begins by examining systems of textual classification because of such questions regarding the boundaries between the scientific and poetic imagination. Given the number of debates surrounding allegory as a symbolic mode, it is impossible to discuss allegory without considering the criteria by which it is defined. *Through the Daemon's Gate: Kepler's Somnium, Medieval Dream Narratives, and the Polysemy of Allegorical Motifs* analyzes the cosmological dream allegory in order to examine the ways that allegory produces meaning. This narrative model, evoked through the journey to the stars, had a particularly strong hold on the late medieval and early modern poetic imagination. We can find parallels to this specific type of journey with the heroic quest of knights of legend and with records of actual travel, ranging from accounts of the new world to journeys across the Holy Lands.

All of these means of evoking travel respond to cultural and philosophical circumstances. The cosmological dream narrative, like other travel narratives, describes physical movement. At the same time, this physical movement evokes the intellectual journeys of writers exploring the forces governing the universe. The early modern period demonstrates a moment of reappraisal regarding these forces. The supernatural informs the scientific, and each approach suggests different cosmic structures.

However, the supernatural and the scientific are not mutually exclusive. For instance, in *Myth and Science in the Twelfth Century: A Study of Bernard Silvester,* Brian Stock points out that Bernardus Silvesteris' ability to map the macrocosmos onto the microcosmos directly results from a materialism and interest in the natural world that developed, paradoxically, out of a Christian

Neoplatonism seemingly concerned only with the transcendent. In the *Cosmographia,* Silvestris, in keeping with the Platonism of the *Timaeus,* depicts an allegory of the creation of man as coordinated by Nature. The division of the book into two sections, Macrocosmos and Microcosmos, represents Silvestris' transcription of antique sources into a vision of the natural intended to reconcile pagan and Christian worldviews. While Silvestris' adherence to Platonism differs considerably from Kepler's cosmography, we cannot view these earlier allegories as monological or rhetorically absolutist. Instead, "It is largely in the tension between what we may call the 'conventional' Platonism grounded in the *Timaeus* and the potentially contradictory implications of these heterogeneous elements in Bernardus' thought that the meaning of the *Cosmographia* consists."[9] The rhetorical complexity of this and other sources, such as Macrobius, used by Kepler, testifies to the significatory aporia enabled by the shared genre and mode of the *Commentary on the Dream of Scipio, Cosmographia,* and *Somnium.*

In fact, Macrobius' *Commentary* and Kepler's *Somnium* serve as the two historical poles of my discussion. We must collate these two texts for better understanding of the extent to which narratology recapitulates cosmology. Indeed, Cicero's *Somnium Scipionis* is listed first among the catalogue (an example of *amplificatio*) of Kepler's literary sources. Cicero's *Somnium Scipionis* was most familiar to medieval audiences as the subject of Macrobius' *Commentary,* which served as an authoritative analysis of both dreams and cosmology regarding the putative dream about the great Roman general originally as expressed by Cicero. Kepler, the protomodern with medieval tendencies, guards the gate separating ancient and modern astronomy. Kepler's role in the medieval tradition cannot be underemphasized. Bruce Stephenson's assertion in *Kepler's Physical Astronomy* that Kepler "completed the structure of ancient astronomy"[10] attests to the significance of Kepler's relationship to medieval philosophy and theology and to classical thought. My study aims to resolve the frequently-evoked dialectical opposition between Kepler's poetic and scientific imagination. The first main section of my argument, consisting of Chapters Two through Five, provides an overview of the generic qualities of the cosmological dream allegory. Recognizing the dynamic quality of literary genre, I discus the transcription of the astronomical details of the Ptolemaic cosmos into the narrative form of cosmic allegories. The second section, containing Chapters Six through Nine, provides a detailed analysis of the posthumously published *Somnium, or Lunar Dream,* of Johannes Kepler. I contend that this narrative, emblematized by the Daemon at the center of a deeply embedded structure, serves as a gateway for the possibilities available to the genre of the fabulous narrative. This Daemon, redolent of the chaos of a universe and a textual form not circumscribed by the holy, attracts and repulses. Kepler uses this genre to articulate something more complex than a dialogue between two world systems. Indeed, the form of the *Somnium,* largely ignored until the twentieth century because of its baroque features, speaks more to the pandaemonic explosion of philosophical, theological, and scientific questions unleashed in the sixteenth and seventeenth centuries than that expressible in a form like the literary dialogue. Compared to the heteroglossia of Kepler's *Somnium,* Galileo's *Dialogue* comes off as distinctly monological.

But, besides merely championing Kepler, or promoting the *Somnium* as yet another worthy text glossed over by a teleological canonizing process at work in both science and literature, I have produced this study as an example of the semiotic analysis of motifs and genres mutated by scientific developments. While Kepler serves as the endpoint of this study, my analysis of the transformation of the *narratio fabulosa* over time emphasizes the recursive relationship between natural and textual worlds continuously (re)writing one another, encysted and encrypted with meaning.

The exterior of the dream, the narrator's waking life, operates as a shell which encases the successive narrative layers. Thus, the outermost layer, the dream, conceals what is, from a narrative standpoint, an innermost layer that consists of the journey to and description of the cosmos. I start with this problem in Chapter Two, "Allegory and Movement." For medieval philosophy, the natural world operates either as a personified figure that speaks like a text or as an unspeaking form that cannot be expressed in a text. In this section, I emphasize the extent to which narratives move between these poles of representability through a discussion of kinetic energy and allegorical structure articulated by Gay Clifford in *The Transformations of Allegory.* I apply her concept of the energy expended by the allegorical hero to the cosmological dream allegory, contrasting it to the more standard model of the terrestrial and quotidian quest narrative.

This discussion of movement and the natural world leads me to a necessary reconsideration of the sublime, a category linked to allegory since Romantic poetics. Using an example from Horace's *Odes* of poetic inspiration as a temporary paralysis of otherwise frenetic movement, I point to the literary dream as a narrative strategy which allows for the embodiment of the divine, and question the relationship between the medieval divine and sublime. From there, I move to a discussion of the dream as both narrative frame and setting. I consider the medieval literary dream as a mode which creates a narrative space constituted by a direct struggle between the body and the spirit. This then leads to a discussion of the contrast between the physical location of the immobile dreamer and the extravagant movements of the dreamer within the dream.

In Chapter Three, "The Limits of Language as a Celestial Vehicle," I provide a discussion of the *narratio fabulosa* from the standpoint of medieval semiology and the representability of the divine. I discuss movement, the main theme of the chapter, as a component of Augustine's description of celestial harmony, a necessary mathematical concept for astronomy, and an important element of the narrative structure of allegory. I begin the chapter with an overview of the linguistic representation of Divinity, using Dante's *Paradiso* as an example leading to the problems of referentiality which distinguish the cosmological dream allegory from quest narratives built from terrestrial topos.

Moving to Augustine's definition of the sign, I discuss the theory of correspondence between macrocosmos and microcosmos, particularly as exemplified by the fusion of philosophy and mathematics that was used by Neoplatonists in support of this theory. As Robert Jordan notes in *Chaucer and the Shape of Creation: The Aesthetic Possibilities of Inorganic Structure,* the Neoplatonists "following the teachings of the Pythagoreans [. . .] envisioned the Creator as a master mathematician."[11] The well-known influence of Plato's *Timaeus* on the Neoplatonic cosmological model presents cosmos as "a detailed mathematical demonstration of God's adherence to Pythagoreanism—which is to say, it demonstrates God's rationality in the quantitative language of pure reason."[12] However, such a critical stance, as I argue, overlooks the significance of language in favor of mathematics. The ostensible excision of mathematics[13] from language, in fact, enacts a hermeneutical process that contributes to the development of the cosmological journey as genre. In other words, the transition from mythic to mathematic language demarcates a quest undertaken from natural philosophy[14] to natural science.

The celestial model, then, constitutes the essentially textual problem of embeddedness. To speak of the cosmos, the cosmologer begins already through the imposition of an additional sphere: that of the cosmology itself. This textual shell encases even the celestial sphere, the outermost shell of the knowable regions of the macrocosmos. Illustrations of the Ptolemaic world system, such as those featured in Peter Apain's *Cosmographia* (1524), emblematize the threshold between the text and the universe. In such illustrations, the "Habitaculum Dei" surrounds the celestial sphere. But, the realm of God is itself delimited by the margins of the page.

Macrobius, an authoritative source for medieval and early modern cosmologists, tells us that the celestial sphere may also be called the "fixed or immovable sphere because the stars appear to be fixed in it."[15] The celestial sphere also frames or encompasses the rest of the universe, a fact which Macrobius demonstrates through his discussion of the astrological significance of numbers. The Christian identification of the Platonist World Soul with the *Logos,* as opposed to its earlier identification with the non-deified and not even necessarily textualized *logos,* implies something more than a multilayered world written by the Word. Instead, the cosmologist, when describing Word and world, embeds these within a text. The transition from Christian philosophy to science, or the attempt to reconcile philosophy and cosmology, must then be examined as textual problems affecting the discourse of writers seeking to confine what cannot be written within writing. The dream as narrative device, I argue, presents a solution to the tricky problem of representing the unrepresentable. Further, the dimensions of this problem correspond to the perceived finitude of the cosmos: the recognition of infinity effectively completes the cosmological allegory as genre.

The corporeal and temporal realm of the Ptolemaic system featured a series of progressively smaller interlocking spheres with a stationary Earth at the center. This cosmology, describing finite space, likewise necessitates an aesthetic valuation of proportion. The *Timaeus* emphasizes the relationship between the "right" proportions of the universe and its inhabitants:

> the god, wishing to make this world most nearly like that intelligible thing which is best and in every way complete, fashioned it as a single visible living creature, containing within itself all living things whose nature is of the same order.[16]

The dreamer, dreaming of a voyage into these spheres, and perhaps beyond the limits of the celestial sphere, would be dreaming of a dreamer at the center of these spheres. However, the narrative of the dreamer begins with a frame that, paradoxically, encases the encasing spheres of the macrocosmos. A rhetorical and textual envelope encapsulates the literal cosmic one, and also acts to further distance the writer of the dream from the cosmos of the dreamer: the writer does not rewrite the cosmos; instead, the writer can testify that the contents of the dream represent a non-textual and non-rhetorically tainted vision. Thus, the universe itself, distant and unknowable, exerting its mysterious influences on the Earth, becomes tangible, and therefore describable, within the clearly delimited space of the dream. The dream would also include a detailed account of the shape of the macrocosmos as the dreamer journeyed further from the Earth, projecting his consciousness and form out into that crucially anthropomorphic or animate nest of spheres.

Given the inherently textual nature of cosmologizing, the question of authorship takes on an anxious significance for the writer who, more like Satan than Dante in the *Inferno,* embeds himself within the text, a sheet of ice, a mirror, an ocean frozen in time, by mapping, and thereby delimiting, cosmos through narrative. The case of Tycho Brahe's introduction of his world system, for example, testifies to the primacy and authority garnered through clearly attributable authorship. Brahe, fearing plagiarism, inserted a brief account of his world system in his book on the comet of 1577 despite the fact that this universe merely represents a modification of Copernican theory intended to serve as a resolution to the theological controversy in Denmark between the Ptolemaic and Copernican systems.[17] The Tychonic system, characterized by an unmoving earth circled by the moon and sun, differed from the Ptolemaic system in that it did away with "that immense structure, the sphere of the fixed stars"[18] and allowed the planets other than the earth to orbit the sun. Brahe's creation of the system as an argument intended to bridge competing factions, paired with his furious reaction when "an avowed enemy of his, Reymers Bär ('Ursus'), who had begun life as a swineherd and had become a professor of mathematics,"[19] allegedly copied the system, implies that celestial correspondence was a two-way process: above all, cosmologists like Brahe desired to create the universe in their own image.

The theory of celestial correspondence, that cosmic proportions determine living forms, relationships, and personalities, justifies, then, a chain of inscription stretching from Logos to cosmos to the individual. In the same way that Brahe feared an accusation of plagiarism that would have prevented him from scratching his own signature into the firmament, the narrator of Chaucer's *The House of Fame* fears stellification, or the process of transcription into the stars: this fear emphasizes the inherently textual components of the dream as vehicle. In Chapter Four, "The Process of Stellification," I use this word *stellify* as a starting point for an examination of the process by which the dream, as a narrative vehicle, inscribes the human onto the cosmos in a manner evoking Northrop Frye's conceptualization of anagogy. By this, I mean to invoke the gap between the dream as a message received from the heavens and the dream narrative as a rhetorical strategy intended to invoke a particular effect in the audience. This chapter is, above all, an overview of dream books and systems of dream classification.

Chaucer parodies the cosmological allegory through the many ostensibly formal invocations to the gods and the muses in *The House of Fame.* However, these allegories were not intended to merely recapitulate the views of classical authorities. Instead, the cosmological dream allegory shows writers progressively writing themselves onto the stars. The movement from Macrobius to Kepler exemplifies a narrative form that became increasingly effective for the presentation of an empirical model of the universe. This is not to say that the cosmological dream allegory was areligious or scientific in our sense of the word. Indeed, my discussion of John of Salisbury's critique of the medieval dream book in Chapter Five points to the differing conceptions of such categories for the medieval imagination. As a composite genre, the dream allegory remains sacred and profane. The fusion of these disparate, and decidedly fantastic, elements helped to promulgate an emerging scientific discourse and consciousness.

Cosmological allegories, from Alain de Lille to Kepler, are concerned, above all, with presenting the shape of the cosmos. The aim of presenting the shape of the universe, however, is as philosophical as it is representational. Hans Robert Jauss comments in *Question and Answer* that "The works of the aforementioned authors [Claudian, Boethius, Bernardus Silvestris, Alain de Lille, and Brunetto Latini] are so permeated by a single 'ultimate question' that they could be read as individual installments in a series, each of which provides a new answer to the problem of the ongoing existence of the world."[20] I would argue that this list should also include Kepler; however, my analysis does not argue for the essentially linear solution of philosophical problems as posed by Jauss. Kepler, in many ways, demonstrates that knowledge does not grow linearly, but recursively. Still, Jauss' assertion for the fundamental differences between these texts helps to support my claims for the polysemy of a form that has often been derided as a featureless shell.

My emphasis on the essentially rhetorical features, the hermeneutic and structural richness, of cosmological allegories builds on a critical trajectory starting with Gerald Holton and Thomas Kuhn and extending to more recent works such as *Metaphor and Analogy in the Sciences,* a collection of essays that address the implementation of metaphor in scientific texts. While the subjects addressed range from the works of Kepler to modern physics, all of these essays are concerned with "the theory of metaphor [as] a theory of the abandonment of semantic convention and of the innovation of representations."[21] Indeed, theories describing the workings of the physical world rely on the maintenance of continuous metaphors. While a theory may make a claim toward modeling or explaining empirical reality, it often requires an enabling trope or metaphor itself suggested by cultural contingencies. As such, my study uses the critical terminology of narratology as practiced by Gerard Genette and Tzvetan Todorov because the *Somnium* is, above all, a story and, furthermore, a story that retells earlier stories. The narratological emphasis of my study counters the reduction of complex texts to a series of single statements that can often occur with a purely rhetorical approach. Beyond merely recapitulating the axes and charts frequently associated with narratology, my analysis also benefits from the free play of the critical imagination championed by theorists such as Michel Serres, Michel de Certeau, and Jacques Derrida. In the same way that, as Fernand Hallyn argues in *The Poetic Structure of the World,* scientific invention cannot be distinguished from poetic imagination, writing exploring science must, in its form, also express the imaginative.

As an extension of metaphor through an entire narrative, allegory plays a crucial role in literary and scientific poetics in that it aims at the creation of an interpretive world. Allegory, a narrative process involved and invested in both masking and unmasking, necessitates inversion or other speaking. As Angus Fletcher argues in *Allegory: The Theory of a Symbolic Mode,* this discursive mode involves the transformation of one entity through another. The continuous metaphors employed in the description of a new scientific theory then rely on concealing metaphor in a structure which can be deciphered or untangled. The Romantic distinction between allegory and symbol demonstrates the extent to which a symbolic approach, because of its universality to its proponents, could not be effectively isolated in a limited generic category. As Jon Whitman points out in *Interpretation and Allegory,* "Romantic theory tended frequently to treat the 'allegorical' as a kind of literature [which was] contrasted with 'symbolic' writing."[22] Peter Szondi discusses a similar problem regarding classification in *Introduction to Literary Hermeneutics.* The development of a "symbolic" writing posits both terms—symbolic and writing (écriture, as derived from Mallarmé)—as universals. As a result, interpretation based on symbolic writing also ignores the extent to which verbal and written transmissions of signifiers create genres. Szondi refers to Friedrich Schleiermacher's work as an example of a "modification of the task of hermeneutics [which] effects its emancipation from the disciplines as whose auxiliary it is usually regarded."[23] This process, dismissed by both Romantics and Modernists, often invoked along with the hermeneutic labors necessary to untangle Christian philosophy, is also suited to the works of early scientists. Figures such as Johannes Kepler build a discourse which, like allegory, "present[s] an aesthetic surface which implies an authoritative, thematic, 'correct' reading, and which attempts to eliminate other possible readings."[24] Of course, given Newton's obsession with allegorical and hermetic interpretation, prominently evident in his unpublished writings, we cannot conclusively claim that the posthumously published *Somnium* serves as the epitaph of the medieval imagination.

In my chapters on Kepler, I address specific motifs that feature prominently in the *Somnium.* Chapter Six, "The Journey, the Book, and the Dream: An Overview of the *Somnium,*" begins with the textual history of the *Somnium,* and explains the importance of this text for interdisciplinary approaches to science and literature. The title of Chapter Six demarcates the three main components, the book, the dream, and the journey, that Ladina Lambert's recent *Imagining the Unimaginable* designates as most integral to the structure of the *Somnium.* I have adopted these three components as sites of inquiry for three of the primary motifs of the work, these being the circle, the moon, and the Daemon. In keeping with the problem of embedding, I contend that the book contains the dream that, in turn, contains the journey in the same way that the circle contains the moon contains the Daemon as allegorical motifs. The relationship between these motifs, each of which undergoes transformation, impels further resignification of each. Thus, the circle becomes an ellipse as the mythic moon becomes a material satellite as the Daemon, a possibly diabolic messenger, becomes the Daemon as ambassador of a universe (and not cosmos) describable without recourse to diabolic messengers.

As a geometric figure, the circle opposes the ellipse in a manner bespeaking greater complexity than the classical/ baroque dichotomy inaugurated by Koyré on this subject. The story of Galileo's attachment to the circle and Kepler's adoption of the ellipse does more than describe the aesthetic leanings of these scientists. The circle emblematizes the circular logic of Galileo's preferred form, a dialogue of personified viewpoints speaking with a predetermined and unified goal. The ellipse, a crushed circle, by contrast, denies, for the medieval imagination, seemingly natural argumentative logic. Chapter Seven, "The Poetic Structure of the Circle," contrasts the circle and the ellipse and the attendant significance of these figures to Galileo and Kepler's theories of motion in light of claims made by Marjorie Nicolson and Arthur Koestler that the *Somnium* enables a new genre. Building on Hallyn's work, I argue that Galileo and Kepler's differing concepts of motion can be conceived of as modal distinctions: their differing interpretations of motion extend to the methods of presentation favored by each.

Chapter Eight, "Kepler's Allegories: The *Somnium* is not a Somnium," analyzes the *Somnium* in relation to the Macrobean dream categories that I first discuss in Chapter Three. In response to arguments made by Fernand Hallyn and Ladina Lambert, I do more than categorize the *Somnium* within Macrobius' schemata. Instead, I examine the interrelation between medieval cataloguing systems for two nighttime phenomena: dreams and the moon. In the same way that medieval cataloguing systems do not allow for a sufficient categorization of the *Somnium* as dream, the body of classical sources does not cover the range of significations available to the moon once it ceases to be a mythic no-place and becomes a physical body.

In my final chapter on Kepler, "The Speech of Daemons," I focus on the Daemon's speech, which constitutes the central core of the elaborately framed narrative. I argue that the Daemon, a polysemic allegorical assemblage of the Christian and scientific imagination, represents Kepler's attempt to resolve competing discourses available for theorizing nature. In this chapter, I use Paxson's discussion of narrative embedding to contend that the asymmetry of the narrative frames allows the Daemon's speech (the innermost frame) to infect and transform the outer frames, thus enabling a new way of "speaking" the natural world. Furthermore, I contend that the Daemon's almost disembodied speech becomes a kind of body. This Daemon belongs to a pantheon of hybridic monsters, "disturbing hybrids whose externally incoherent bodies resist attempts to include them in any systematic structuration."[25] Cohen's assertion that "the monster is dangerous, a form suspended between forms that threatens to smash distinctions" characterizes the complex role of this Daemon at the intersection of theology, philosophy, and science.[26]

In this study, I provide a case study for the interpellation of allegory, history, and realism. While my examples take a medieval/ early modern trajectory, I intend my conclusions to contribute to current discussions regarding allegory in contemporary literature and science, like Theresa Kelley's argument in *Reinventing Allegory* that allegory "is a historically contingent genre and idea whose survival in modernity retrospectively conveys the cultural and literary interest of its earlier forms and historical moments."[27] Before examining specific examples of cosmological journeys, I intend to discuss how these texts operate as a genre created as a response to the natural world. The critical viewpoint a writer uses to approach allegory determines its range. The Ciceronian conception of the ability of allegory to affect discourse differs from that postulated by poststructural theorists. Much rhetorical theory, for instance, clearly relegates allegory to the subordinate position of rhetorical figure of speech. As a trope, it is a tool, an extended metaphor, capable of being used in any discursive practice. This ahistorical approach implies that the process of allegorization inheres in the deep structure of language. Furthermore, such a standpoint subordinates the importance of the use of allegory within a given text. The critical reader of an allegorical text may catalogue the devices used, but, through such an approach, the subject of the allegory supercedes the generation of that allegory. Allegory illuminates a writer's subject, but this subject occupies a more prominent position than the method used to approach that subject. For the rhetorician, allegory must be a stable concept: the mechanics of its applications do not need to be investigated. Instead, readers should be interested in the effect generated by its application.

The title of this book signifies the confluence of narrative genre, allegorical motif, and scientific concept that I evoke in my study. The Daemon's gate signifies as a textual and conceptual opening. The multiple layers of the *Somnium* coalesce or collapse at this gate. By "layers," I mean the various levels of this embedded narrative, the competing generic categories evoked by the work, as well as the possible significations of the Daemon as a learned expert or diabolical manifestation. My analysis demonstrates that the change in ideas precipitated by the new astronomy extends to the narrative forms used to express these ideas. At the same time, we must renew our investigation of the connection between scientific knowledge and spirituality. Rather than viewing the rejection of geocentrism as a linear movement from outmoded theologically-influenced models of the universe to a factually correct model divorced from the divine, we must, instead, recognize the importance of spirituality in the development of the scientific. Likewise, this investigation must move beyond a narrow cataloguing of doctrine; elements such as literary genre and the dominant motifs in a work, for instance, are often themselves indicative of the competing ideologies influencing the formulation, expression, and articulation of scientific concepts.

Chapter Two

Allegory and Movement

The astronomer contemplates the night sky. What is the scale of this image? How do we imagine his contemplation? On one hand, perhaps we see the astronomer as a tiny shade perched on a hill. Above, the sky stretches out like the limitless ocean. Below, the astronomer shares the countryside with flocks of sheep nestled together against the cold wind, clusters of fruit trees, the distant lights of a provincial village.

Or, we picture a different scale. A different astronomer, perhaps. One on equal footing with the stars. Equipped with parchment, telescope, and astrolabe, the astronomer rises into the heavens. He fills the heavens with his measurements. He names the stars, corrals their infinitude into constellations. The stars, after all, can only repeat the stories first told by the fire in the distant village.

The scale of the image determines the shape of the cosmological journey. A journey into infinity surpasses movement. It overtakes measurement. The astronomer's tools vanish in the void. In a finite universe, our astronomer is an architect. The stars glimmer like the separate stained-glass panels in an immense cathedral.

ALLEGORY AND MOVEMENT: THE NATURAL SUBLIME

The problem of the sublime, based as it is on the estimation of scale and movement, has a direct bearing on our attempt to outline the cosmological dream allegory, and to see how this narrative mode contributes to the creation of scientific discourse. This abstract aesthetic problem, as my example points out, cannot be confined to the woodland frolics of Romantic poets. Indeed, as Wolfgang Kluxen points out in "Nature in the Ethics of the Middle Ages," the contemporary conception of nature and the natural has its roots in the medieval commingling of the natural and natural law, so that the rediscovery in the twelfth century of "a rationality founded on nothing else and in need of no justification"[1] sets the stage for the division of rational knowledge and sensual experience characteristic of the Romantic definition of the sublime.

While commentators have customarily denied the feeling of sublimity to medieval audiences, critics such as Piero Boitani have argued for a reevaluation of the sublime and medieval audiences. This problem, related to the cultural perception of the natural, becomes increasingly important for modern scholars interested in the history of nature as experienced, conceptualized, and allegorized. The writing of a cosmos involves an imaginative procession through its interior; the writing of a universe entails a mapping of a limitless expanse. The distinction between cosmos and universe, contingent on, as Koyré elucidates in *Galileo Studies,* (in)finitude, then, weighs heavily on the narrative representation of limited or unlimited space.

Johannes Kepler, his significance measured by "the intensity of his contradictions, and the use he made of them,"[2] plays the role of both astronomers in my example at the beginning of this chapter. Perhaps, in keeping with Arthur Koestler's evocation of the power of Kepler's "sleepwalking unconscious self," the first astronomer dreams the second astronomer, or vice versa. Of course, like the second astronomer, Kepler clings to the finitude of the solar system. His work enables infinity of the physical universe, but he himself does not contemplate its limitlessness. Beyond the solar system lies the unknowable void, lies divinity.

Indeed, the dimensions of this scale affect our ability to discuss the sublime in the cosmological dream allegory. The sublime, a quality associated with the poetic inspiration of the natural, remains problematic when applied to medieval texts. However, this controversy necessitates an exploration of the topic, which becomes increasingly relevant as scholars, following the example of Pierre Duhem, seek "to demonstrate that the philosophy of nature and the cosmological theories developed in the late Middle Ages prepared the way for the new sciences."[3] Used in the service of early modern science, the medieval dream narrative, a form predicated on the meditation of the spatial construction of a finite cosmos, impels recognition of the infinity of nature, a key component of the modern sublime. In this chapter, I am concerned, then, with the feeling of the sublime, with its various definitions or interpretations, and the spatial dimensions of the journey of the cosmological traveler.

The sublime as a concept calls up estimations of mathematical vastness or extremes. Kant, building from and commenting on Edmund Burke's *A Philosophical Inquiry into the Origin of Our Ideas of the Sublime and the Beautiful,* connects beauty to the finite and the sublime to the infinite. The second book of his *Critique of Judgment* elaborates his separation of the sublime into mathematical and dynamical categories. The vantage point one takes when conceptualizing the order of the cosmos, or the infinity of the universe, determines one's perception of beauty or the sublime. Based on Kant and Burke's differing takes on the sublime, scholars have tended to situate the sublime as a specifically Romantic, post-Enlightenment concept. However, as Burke points out, the Romantic formulation of this concept is clearly indebted to Longinus' *On the Sublime.* The Romantic sublime, defined by Thomas Weiskel as the idea that "man can, in feeling and in speech, transcend the human" is, of course, experienceable in a pre-Romantic world. In fact, such a definition of sublimity could be applied, with equal precision, to the visions of religious mystics. The exact nature of the confrontation between the human mind and natural phenomena producing the effect of the sublime, however, necessitates fuller historicization. While a medieval sublime may differ qualitatively from the concept popularized by Romantics, an analysis of this concept elucidates the influence of the natural on the narrative forms of cosmologists. Longinus' *On the Sublime,* importantly, shows that the sublime is not a strictly modern concept. The presence or absence of the sublime is linked to the philosophical conditions and aesthetic preoccupations of different eras. C.S. Lewis, for instance, denies the sublime as an aesthetic category available to the medieval imagination. He centralizes, instead, beauty and the sacramental symbol.

In *The Discarded Image,* Lewis employs exactly this image of the contemplation of the night sky to articulate the primary distinction between the medieval and modern conceptualization of the infinite. He remarks that "to look out on the night sky with modern eyes is like looking out over a sea that fades away into mist [. . .] to look up at the towering medieval universe is much more like looking at a great building."[4] These two images suggest dual possibilities for the contemplation of the shape of the universe. Lewis' evocation of the modern universe as a limitless ocean suggests a space that contains no single point from which to observe the entirety. But, the architectural metaphor of the universe as building places the medieval universe within the realm of humanly-scaled geometries.[5]

The medieval cosmos can thus be both finite and tactile, an object as tangible as a physical body or a great temple. This quality, expressive of the fundamental components of Aristotelian physics, "represents a purely static conception of order."[6] This statement, from Koyré's *Galileo Studies,* testifies to the dependence of the concept *cosmos* on purely physical concepts. Indeed, cosmological diagrams such as those found in Thomas Digges' *A Perfit Description of the Coelestial Orbes* (1576) represent this static conception of order in architectural form. The cathedral, solid, monumental, and unmoving, represents the cosmos. Built from interlocking rings, the universe testifies to the clearly demonstrable order of the divine intelligence, and the static perfection of finished form. In Digges' book, he even refers to the fixed sphere in architectural terms, calling it a "pallace."

The ancient model of interlocking spheres enables the universe as an "object in which the mind can rest, overwhelming in its greatness but satisfying in its harmony."[7] Kepler's *Mysterium Cosmographicum,* for instance, argues for a geometric conception of the universe, or of the universe as structure. The *Astronomia nova,* on the other hand, relies on the belief in a Divine harmony underlying a universe the shape of which no longer corresponds to the structural ideal presupposed by the architectural metaphor.

To support his claim that the medieval mind was unable to experience sublimity, Lewis provides Dante as an example. Returning to his architectural analogy, Lewis notes that "Dante is like a man being conducted through an immense cathedral, not like one lost in a shoreless sea."[8] The opposition between cathedral and ocean stems from the opposition between finite and infinite. John M. Steadman supports this customary view. Even the poetic forms used to evoke the natural rely on formulaic and conventional topoi, so that "Nature thus becomes a kind of Aristotelian telescope [. . .] where the poet contemplates the universal in the particular and the 'concept' in the 'percept.'"[9]

Not all writers reject the possibility of medieval sublimity. The cathedral, like the Neoplatonist model of the universe, evokes the divine universe within its towering confines. To this extent, the cathedral can express contemplation of the infinity of the Divine. Piero Boitani so argues this point in his refutation of Lewis' attitude toward the medieval sublime. Boitani points out the crucial passage from the *Convivio* that Lewis ignores in his estimation of the medieval capacity for the sublime. In this passage, found in *Convivio* IV, Dante defines awe as "uno stordimento d'animo" ["a certain bewilderment of the mind"] caused by an encounter with "grandi e maravigliose cose" ["great and wonderful things"]. For Dante, worldly phenomena evoke the sublime feeling of awe. "Maravigliose cose" provoke this bewilderment of the mind. Even miracles are expressions of divinity through natural phenomena. The sublime effectively paralyzes the finite organism before the recognition of infinity. Boitani recognizes divinity as the key factor distinguishing Dante's evocation of the sublime from that employed by eighteenth century writers. Still, the contemplation of the infinite and the recognition, for the medieval mind, that the sublime is a feeling of spiritual reverence, does not invalidate such an experience as a confrontation with the sublime.

Lewis' argument that the medieval mind was unable to experience the sublime, however, portends more when we consider the sublime as a category within an interpretive framework. Romantic poetics, by no coincidence, depends on the imaging and theorization of the sublime. The connection comes not merely from the Romantic preoccupation with the natural as a source for poetic motifs. It also stems from Romantic theories of artistic signification. The symbol, defined by the Romantics as a free-floating signifier clearly differentiable from the stable and finite hierarchy of allegorical signification, suggests infinitude.

Allegory, as an interpretive system, parallels the model of the medieval universe that Lewis provides. Based on finitude, both are structures with discernible dimensions. Allegory may rely on a finite set of interpretive possibilities. However, like the Neoplatonist model of the universe, Christian allegory in particular denotes the existence of the Divine. For Lewis, this denotation of the Divine does not itself express Divinity. Further, the finitude of allegorical systems prohibits a contemplation of infinity. Both the model of the universe and the interpretive framework of Christian allegory are built from an interlocking, but limited, series of literal and figurative shells: the spheres and the levels of allegorical meaning. However, to return to Boitani's critique of Lewis, the strength of spiritual revelation uncovers a new dimension of meaning, or brings about a greater involvement of the viewer with the spiritual. Thus, "the expression 'stordimento d'animo' (literally, the mind's 'stupefaction' or 'daze') is stronger than anything Aristotle ever used, and in reading it one is reminded of Longinus' coupling of the marvelous (*to thaumasion*) with *ekplexis,* the shock of amazement it produces."[10] The term *ekplexis* describes the reactions of the cosmological traveler. Boitani likens *ekplexis* to the shock of amazement produced by the contemplation of the marvelous. Linked to *ecstase,* it also evokes a state of spiritual ecstasy, like the sublime rapture of a blood-besotted maenad. Such a rapture could equally explode in a Barthian *jouissance* of the text. For medieval philosophy, indeed, the physical world was a text written by an energetic and joyful Word.

The finite system of possibilities engendered by this allegorical text still allows for divine revelation. Such a constant presence of divinity serves to produce this shock of amazement through natural phenomena. The supernova of 1604, for example, led to a great number of texts which speculated about its significance. Interpretations varied, and ranged from the arrival of a global conflagration to the rebirth of Christ.[11] Kepler's *De Stella Nova* also provides an astrological interpretation of the nova: "Kepler naturally regarded it as significant that the new star should have appeared at the time of a conjunction of Jupiter, Mars and Saturn and in a position close to that of the conjunction."[12] The nova provides an example of a natural event viewed as a textual signifier. Furthermore, it is a new signifier: the star was invisible to the naked eye before it exploded.

A new event in nature, like a new interpretation of a pre-existing natural phenomenon, widens the set of possibilities available to the book of nature. This is particularly important given the interconnection of the text as universe and the universe as text. The development of a modern science comes from a reevaluation of the book of nature. This activity was not separate from Biblical interpretation. While the book of nature was previously thought to completely reflect scripture, "the difficult point in the [re]thinking resulted from the supernatural in nature."[13] These phenomena that cast doubt on the preceding model of the universe impact the connection between Scripture and nature. However, Scripture necessitates the interpretation of nature: "together with the disclosure of God by the word came the disclosure of God in the work; together with the book of Holy Scriptures came the book of nature, the explanation of which would be considered man's great duty."[14] This discrepancy is particularly strong in Kepler, writing between multiple sets of interpretive possibility. Most specifically, the recognition of a finite or infinite universe affects the possibilities available to Kepler. While working within the context of the finite universe, he makes discoveries which point to the infinite universe.

The sublime as an aesthetic quality effects the operational model of the universe. Allegory as well invests this process with interpretive plenitude. The finitude of the universe then provides a point of distinction not just for the quality of modernity, but also for the interpretations available to natural phenomena. Thus, in *From the Closed World to the Infinite Universe,* Alexandre Koyré argues that Kepler bases his adherence to a finite model of the universe on theology. Kepler values the finitude of the universe because a finite universe more closely corresponds to the interpretive possibilities of Christianity, so that "Kepler sees the Sun, the stars and the intervening aether as manifesting the Trinity of the Father, the Son and the Holy Ghost."[15] The finitude of the universe, above all, expresses harmony, a key component of the Neoplatonist model. At the same time, the contrast between theological and scientific interpretive systems expresses a simultaneous "desire to break away from theological restraints"[16] and "a radical change in the understanding of symbols and in the methodology of interpretation"[17] commonly associated with the Renaissance.

The shape of the physical universe and the numerical possibilities of Biblical typology correspond to produce this harmony. Thus, "this analogy would clearly be entirely unsatisfactory if the Sun were no more than a typical star"[18] or if the earth did not hold a privileged position in relation to that star. Kepler had located the earth as the central planet of the solar system, making it the ideal position from which to regard the universe. This is just as well, because the dimensions of the universe can only be explored from an earth-bound observation point. This single point of observation plagues Kepler's ability to accurately verify his theories. The earth might offer an ideal vantage point, but the single vantage point is insufficient to verify the shape of the whole universe.

This ability to imagine beyond the earth informs early modern astronomy and shapes the allegorical practices used to represent the universe. As Bruce Clarke argues in his discussion of allegory and motion in *Energy Forms,* "The allegorical depiction of a cosmos or encoding of a text both constructs a series of literal or figurative equivalences and provides or implies multiple frameworks within which to read those systems of signs."[19] Physical movement, most importantly, helps to enable allegorical signification. This point importantly speaks to the relationship between cosmography and narratology: neither the cosmological narrative nor the cosmos it represents exist as static forms. Instead, movement through a narrative (the acts of writing and reading) and the cognitive movement or displacement (used to great effect by Kepler in order to figure the orbit of Mars) necessary to imagine a cosmos inform one another.

Gay Clifford's *The Transformations of Allegory* includes an intriguing aside that illuminates this connection between physical movement and the narrative structure of allegorical texts. While she is primarily concerned with quest narratives, her comments also help to explain cosmological narratives that lead up to and include Kepler's *Somnium.* While she is concerned with a general anatomy of allegory as a literary mode, this concept suggests the popularity of the celestial journey as a motif. Clifford parallels kinetic energy, defined as the power of doing work possessed by a moving body by virtue of its motion, and the general motif of the journey as "the fundamental narrative form of allegory."[20] Clifford stresses the extent to which allegory, as a mode, involves movement and process in a way unavailable to the relatively static narrative form concerned with individual symbols:

> The strength but also the limitation of symbols is that they tend to be static [. . .] In allegory the concern is always with process, with the way in which various elements of an imaginative or intellectual system interact, and with the effect of this system or structure on and within individuals. To express change and process allegorical action often takes the form of a journey, quest, or a pursuit.[21]

For Clifford, kinetics, or the branch of mechanics which investigates the relations between the motions of bodies and the forces acting upon them, operates as an explanation for the interaction between the allegorical protagonist and the goal of the quest.

Thus, "we interpret the significance of the 'motion' of the characters and of the forces affecting them in the light of knowledge about the direction in which they travel."[22] This analogy between physical motion and narrative structure allows Clifford to stipulate physical motion as a significant aim of narrative, a notion that Bruce Clarke has reexamined as the "transit to obsolescence" so characteristic of allegory in the service of fashion.[23] Further, the allegorical protagonist intuits both this goal and the direction of the destination. The examples Clifford provides, from *Piers Plowman* and *The Faerie Queene,* begin with the protagonists each posing direct questions about the direction of the quest, so that, in the example of *The Faerie Queene,* the Knight, lady, and dwarf "prick [. . .]on the plaine"[24] until the "blustring storme" forces them to "finde that path, which first was showne." The direct question itself is not a necessary condition for this kinetic relation between protagonist and quest, however, as "where we do not at once know what that task is, other clues are provided."[25]

Medieval conceptions of physical processes such as the creation of energy and motion owed much to a union of Platonism and Stoicism. As in Calcidius' attempts to reconcile these seemingly contradictory sources in his discussion of the origin of dreams, medieval cosmologists like Bernardus Silvestris likened stoic conceptions of elemental change to Neoplatonist theories of motion and time.[26] Thus, physical movement, as a component of energy, parallels quests which are essentially philosophical in character. In such narratives, though, the movement of the quest supercedes accurate depictions of the natural world.

The cosmological dream allegory, while clearly reliant on the quest as a prime factor which drives the narrative, also reconciles physical observation and philosophical speculation. As a result, the physical universe, the setting of the quest, gains in importance. Secondly, as cosmological dream allegories frequently take the form of frame narratives, both movement and space are constituted within the dream. Or, more importantly, movement and space in the dream are represented in relation to the relative stasis of the dreamer. This stasis is closely linked to the active energy of the dream as a psychological experience and as a literary artifact. The non-movement which itself preempts the imaginative moment of the dream is an act of forgetting, an action which paralyzes the actor. This set of criteria is "itself a condition of a particular exaggeration of style or hyperbolical figuration that tradition has called the Sublime."[27] But the process by which the dreamer/poet approaches the sublime is determined by his/her spatial position. Classical conceptions of the rhetorical and the natural sublime help to illustrate what I mean by the dream itself as a landscape. Indeed, the dreamer encounters a seemingly tangible landscape within the allegorical frame. However, this sublime landscape, as recounted in an allegorical narrative, appears to be both rhetorical and natural. As Piet Schrijvers points out in "Longinus and Quintilanus: The Classical Sublime," these categories are neither separate nor interchangeable.[28]

Authors of dream books and cosmological allegories accord the transcendent a paramount position within an otherwise static and diagrammable cosmos, itself a structure wrapped, like Christo's Reichstag, with the luminous robes of Logos. Schrijvers' argument about the sublime illuminates our understanding of the cosmological allegory as a quest through nature. It is not only that the cosmological allegory relies on the conceit of the dream to move towards a representation of the shape of the cosmos. Instead, the rhetorical selection of the cosmos as a motif denoting both the infinite and a knowable, organizable order parallels the medieval conception of the dream as a space where the material things of the natural world are combined into narratives with a naturally determined, divinely ordered meaning.

Most importantly, Schrijvers calls attention to Horace's depiction of poetic inspiration as a Maenad gazing on snow[29] as a particularly strong example of the natural sublime. The Maenad, the "mad" or "demented" follower of Dionysus who roamed the mountains in the grip of ecstatic possession, was both human and divine. Indeed, the daemonic, an "intervening stage between the human and the divine," "usually marks the climax or Sublime crisis point of the strong poem."[30] While the Maenad was typically invoked, like the daemon, as a messenger or agent of divine action, the Maenad here gazes at a landscape without taking action (the Maenads were believed to tear animals and men into pieces: hence Orpheus' demise). The Maenad gazes on the landscape with a feeling of both rapture and fulfillment, best described as a *fascinens ad tremendum*.

However, in such an evocation, the Maenad remains an external object. To this extent, the feeling of the sublime produced by a work is like the gaze of the Maenad. But we must extend this analogy to further understand the problem of the natural sublime. In fact, our consideration of the Maenad as a figure between the human and the divine illuminates the key role of the sublime in the scientific imagination. Thus, we must distinguish between gazing upon the gaze of the Maenad and identifying or internalizing this gaze as the actual feeling of the sublime. It is one thing to gaze upon the Maenad from an external standpoint. A writer's work, in Horace, produces the feeling that one would imagine is felt by a Maenad gazing on the snow. In this example, the actual mental state of the Maenad is not taken into consideration. Instead, the gaze of the Maenad signifies from an external standpoint. The reader is aware that the Maenad stands for a feeling of intense and ecstatic agitation. The analogy receives its strength from an external consideration of this resolution of two forces; the grandeur of the natural landscape temporarily abates the frenzied actions of the Maenad.

THE DREAM NARRATIVE AS PLACE

The work of art can approximate this feeling of the Maenad, but cannot seek to recreate it in its entirety. The dreamer's lack of physical movement, on the other hand, more closely approximates this tension. Especially because, in the dream, the dreamer is moving. The dream state more effectively produces the taming of the Maenad, of the untamable forces of the unconscious, mentioned in this analogy. Thus, the dreamer, like the Maenad, lays restrained by the force of images. In this way, the dream itself is like a place, a sublime landscape, that produces this paralysis. Further, the places that the medieval dreamer could explore were, while mysterious and fraught with supernatural peril, surprisingly well-mapped. Their contours were codified and catalogued; indeed, the complexities of the dream were not conceived of as the output of single dreamers.

From a textual standpoint, the world of the dream is remarkably similar to the world of the allegory. Clifford comments that, like dreams, "the worlds of allegory are only half-familiar and they are rarely safe."[31] Her description implies that both readers and characters are placed in the allegory as though it were a landscape. In this landscape, Aneither protagonists nor readers can predict with any security what phenomena they will encounter or precisely what these phenomena will signify."[32] Her use of the term 'phenomena' here suggests the extent to which the allegorical text, composed of interpretive frameworks of interdependent signifiers, exists as a place that impacts readers and characters in a manner that is almost physical. Like the Maenad, incapacitated by images, "we have to immerse ourselves in the world of each allegory until we discover its peculiar and persuasive internal logic."[33] The language she uses here is particularly relevant for the connections I am drawing between medieval dreaming and medieval allegories.

James J. Paxson summarizes the medieval attitude toward dreams as follows: "dreams are messages or texts transmitted to specific physical loci rather than to specific individuals. They are not unlike letters mailed through the post to predetermined places. The medieval literary dream undeniably has the status of a narrative artifact."[34] The trustworthiness of the dream landscape, or of the images revealed in the dream, depends entirely on the exterior conditions affecting the dream. Thus, the dream produced solely by physical detriments (caused by an excess of food, alcohol, or both—the medieval *visum* or *insomnium*) could not be reliable. To this extent, the space represented in such a dream could also not be seen as reliable or reflective of outside reality. Paradoxically, an overt influence of the physical body of the dreamer could result in an inaccurate representation of the physical world inhabited by the drunk and/or indigestion-plagued dreamer who we can perhaps envision as an extra in a Brueghel painting. On the other hand, the dream sent from the Divine gains its significance precisely because it is not impacted by the body. This differentiation would, of course, affect the truth value of any dream. However, as the cosmological dream is concerned so much with the description and observation of space, this condition of medieval dreams grows in significance.

The true dream is true because unaffected by physical contingencies. While this makes sense in the case of physical ailments, this also impacts physical observation itself in the formulation of space, the cosmos, as a narrative space accurately represented in the frame of the dream. Thus, the true representation of space comes from the Divine, setting it against physical observation. Of course, the dreamer does not actually observe when sleeping. Still, this does establish a distinction between the cosmos as observed by imperfect human senses and the shape of the cosmos as revealed by God. Further, the cosmos itself, as place, was more closely aligned with the Divine than the physical. Though medieval dream travelers experience the cosmos as physical space, it is a physical space marked by the spiritual and Divine in ways unavailable to terrestrial settings. In this sense, the moon as narrative setting surpasses even the sights of the Holy Land as a supernatural space. Early modern astronomers likewise sought to counter failed human observation through the implementation of mathematics as a language which, like the Divinely sent dream, was sufficiently removed from the physical form to reveal truth.

The content and reliability of the dream rely on the condition from which the dreamer "sees" the dream. Mieke Bal's analysis of setting and subject in *Narratology: Introduction to the Theory of Literature* allows us to understand the narrative complexity of this relationship. In discussing narrative location, Bal refers to "the physical, mathematically measurable shape of spatial dimensions."[35] Bal does not mean that such places are real; she affirms their existence as imaginative constructions within the context of the narrative. However, for Bal such dimensions must exist so that these places can be imagined as representations of real or possible settings. In the cosmological dream narrative, the cosmos itself becomes the place of the narrative. Its form as depicted through the celestial voyage renders it the setting of the internal narrative frame at the same time that the actual cosmos embeds the world of the dreamer. However, Bal's assertion that place must be able to be described conflicts with the essentially non-describable actuality of the cosmos as its representability is conditioned by medieval sign theory. Further, Bal's language in this definition of place is particularly evocative given the relation between figural language and mathematic figures in the usefulness of each for representing the cosmos. The precision of description necessary for a setting to exist as a place is closely related in the celestial dream allegory to this question of language and its limits as an apparatus for representing the Divine (or the mathematical sublime).

In discussing general narrative setting, Bal further qualifies the extent to which setting and character exist in a reciprocal relationship. Thus, place evokes the objective dimensions of setting. These dimensions may or may not be actually conveyed by a character or specifically mentioned in the narrative. In this sense, details of place are suggested but not fully disclosed in the narrative. The concept of space as articulated by Bal relies on setting as depicted in the narrative. Thus, for Bal, places that are "linked to certain points of perception" and "seen in relation [. . .] to their perception" are designated by the term space.[36] The prime distinction between place and space comes from this determination of the function of setting. However, the character immersed in the dream narrative does not actually occupy setting. Instead, Bal's definition of space emphasizes the extent to which the dreamer's presentation of the dream space relies on the conditions surrounding the act of dreaming.

Further, the interpretations available of the dream as space rely on interpretation of the dreamer. To this extent, the dream narrative is, quite literally, embedded *within the dreamer*. The physical and psychological conditions of the dreamer have the potential to shape the dream. However, this conditionality is closely linked to the truth value of the dream. Bal's narratology asserts that narrative necessitates the concept of space. Even if there is no actual character through whom observation is filtered, "an anonymous point of perception may [. . .] dominate the presentation of certain places."[37]

Observation never originates from the consciousness of an actual character: this component of Bal's discussion of space illuminates the disjunction between the dreamer and the setting of the dream in the cosmological dream allegory. The perception of the character impacts the narrative space. It is, Bal contends, impossible to have a narrative setting which does not, in some way, have the subjective signature of a narrating consciousness imprinted upon it. However, the successful dream narrative is exactly that which is not impacted by the obsequious signature of the physical dreamer. In the context of medieval sign theory, the transcendence of the cosmos is only approachable by a narrating consciousness fully removed from a subjective point of perception. The cosmic vision, in order to achieve its purpose, must be clothed in language which depicts a setting as natural without recourse to language rooted in physical reality.

In the earlier example, the Maenad does not actually create the sublime landscape. The same may be said of medieval conceptions of the dream. The dreamer, distanced from the needs of the body, was receptive to dreams. We note the distinction between the dream as something that is lived versus something that is seen from an external standpoint. As Francis Xavier Newman remarks in *Somnium: Medieval Theories of Dreaming and the Form of Vision Poetry,* "In Homer, the word *oneiros* does not mean 'dream-experience' but 'dream-figure.'"[38] Thus, the dream is an object separate from the dreamer. The dreamer may be said to meet the dream. Newman goes on to state that "dreams, in classical myth, are persons, and thus 'real' but they are also message-bearers, and thus meaningful."[39]

This differs markedly from our own conceptions of dreams. Of course people make dreams. They make them, depending on your psychological persuasion, from random events of the day, past traumas, or any variety of informational input. However, this division constitutes one of the key differences between modern and medieval conceptions of both the dream and the dreamer. As S.R.F. Price notes in "The Future of Dreams: From Freud to Artemidorus," "although ancient dream theory was predominantly concerned with prediction, modern interpreters are characteristically concerned with the personality of the dreamer." In other words, a modern interpreter may see the individual dreamer as a repository of potential signifiers. This set of signifiers is either closed or open: it is either made up of a series of personal symbols that can be decoded to give meaning or it features a series of shifting and unstable signifiers that may or may not be attached to any given meaning. For the medieval dream interpreter, however, the images in dreams are like natural phenomena. They can be organized and understood. The meanings of symbols in dreams, like the same symbols in allegorical texts, were codified and hierarchized.

Still, this does not mean that physiological and psychological complexities were entirely foreign to medieval dream interpreters. Indeed, as Salomon Resnik comments in *La mise en scène du rêve,* "Freud reprend, d'une certaine manière, l'idée de la vocation de l'oniromancien qui éclaire la conscience sur ce qui se cache dans l'ombre de l'inconscient" [Freud reawakens, in a certain manner, the idea of the vocation of an oneiromancer who clears the conscious mind of that which is hidden in the shadow of the unconscious].[40] A modern psychologist, for instance, would be familiar with medieval dream interpreters' concern with the age and mental condition of the dreamer. Artemidorus distinguishes between the meaning of a figure in an old and young man's dream, as Robert Levine points out in "Gower as Gerontion: Oneiric Autobiography in the *Confessio Amantis.*" Albertus Magnus, in his *Opera Omnia,* clarifies that the dreamer's *complexia* gives meaning to the dream. The *complexia,* a term that may be translated as the complete psychological profile of a dreamer, suggests an attention to the individual mind that we may more readily attribute to modern psychoanalysis. As Robert J. White indicates in his introduction to Artemidorus' *The Interpretation of Dreams,* "In a sense, Freud, Jung, and others were not so much innovators as restorers, since they were reassigning to dreams and dream-readings the importance that they had held in antiquity, and which they had lost in more recent centuries."[41] A modern concern with the psychological composition of the dreamer informs Artemidorus and Albertus Magnus. Likewise, we should be aware of the suffering of the dreamer: in this context, "suffering" implies the circumstances of the dreamer's life. However, we are also advised by Albertus Magnus to consider the place where the dream occurs as well as the condition of celestial affairs at the time of dreaming. While the psychological make-up of the individual dreamer does impact the dream, the seasons and constellations could also adversely impact a dream, presumably even a dream about the constellations themselves.

The movement of the dreamer is essentially a movement between competing interpretive strategies: interpretation of the dream, of the physical forms encountered in the dream, and the divine significance of these forms. The medieval dream as a textual artifact relies on the synthesis of opposing (even competing) strategies of interpretation and approaches to language. The narrative dream enables motion embedded in the stasis of the dreamer's physical body. Furthermore, these opposing and embedded narrative elements do not conflict in the cosmological dream. Instead, the journey of the dreamer occurs in a space that is readable through the interpretive strategies of the philosophers but maintains the mystery of the supernatural. As such, it is the genre which most clearly allows for an examination of the unknowable Divine language which forms the knowable physical universe.

Chapter Three

Language and Its Limits as a Celestial Vehicle

REPRESENTING THE DIVINE

Representing the divine implies contradiction, and the tacit acknowledgment of the insufficiency of one's own representation. Such an attempt can also be viewed simultaneously as an act of the greatest humility and vanity. An attempt to embody in language that which is disembodied may signal the writer's impertinence and thus cheapen his/her subject. Correspondingly, the writer's desire to represent divinity in human language may serve as an act of the greatest devotion. Dante's *Paradiso,* perhaps the most famous representation of the unrepresentable, seeks to transcend the boundaries of language through language itself.[1] Furthermore, it does this in the vernacular, transcending boundaries of literary convention from the outset of Dante's spiritual journey.

The representation of the divine in Dante relies heavily on process; we can not fully experience the *Commedia* without recognizing the transcendent qualities of movement. A deeper understanding of the *Commedia* necessitates an identification of the physical movement of the journey with the equally physical movement of the pen. On the other hand, these two kinds of movement, so synchronized through the rest of the text, diverge in the *Paradiso.* In fact, while much criticism of the *Paradiso* emphasizes this problem of representation, the evaluation of the connection between writing and movement depends on a critic's evaluation of both actions, writing and movement, as essentially spiritual endeavors. To this extent, Piero Boitani's *The Tragic and the Sublime in Medieval Literature,* for example, emphasizes Dante's encounter with "the supreme light that his memory and his pen are trying to recreate for us."[2] Boitani asserts this point through his analysis of the textual resonance of Canto XXXIII of the *Paradiso,* when Dante "is left mute and alone with his God."[3] Dante's mute condition in this canto communicates the divine mystery revealed to those who write beyond writing, or move beyond movement. Thus, while Dante moves through recognizable physical landscapes in the *Inferno* and *Purgatorio,* in the *Paradiso* he moves beyond the physical realm, discursively and cognitively traveling beyond the usefulness of linguistic referentiality.

On the other hand, in describing the Russian Formalist method as it applies to the *Paradiso,* Fredric Jameson makes the following statement about Dante's approach to representation:

> The formal problem which Dante faces in *Paradiso* is in other words that of telling the story of the timeless in time, of recounting identity in the language of difference, of allowing unity to come to voice through multiplicity. [. . .] Even while Dante the character interrogates the order of paradise and attempts to understand how it can have gradations, Dante the poet continues his poem and carries it forward. We may therefore say that the content of *Paradiso* turns out to be a series of investigations of how paradise could have content; that the events of the poem are 'nothing more' than a series of dramatizations of the pre-conditions necessary for such events to be conceivable in the first place. The subject of the poem is its own coming into being.[4]

His articulation of the contrast between content and form, intended to demonstrate the Formalist approach, denies the content of the poem. There is no paradise other than the paradise coming into being through the act of writing. Such a reading could appear to prohibit or invalidate a theological reading of the poem, built from the aim of "translating all such proposed content back into projections of the form."[5] This problem seems to hinge on whether or not the text is viewed as a discrete unit. From the standpoint offered by an anagogical method of interpretation, for example, Dante's struggle with language in the poem is not encapsulated within it. Instead, these issues are likewise reflected in the natural world. The pursuit of Dantean allegory, therefore, cannot end at the edge of the page, and must proceed by resisting or supplementing mere formalism. Yet a narratology of the new or the ineffable springs from careful understanding of the celestial traveler's rhetorical and figural devices. The writing of a Keplerian new science of the stars and the Dantean pre-figuration of this writing in the guise of the paradisal voyage represent the limits of allegory's narrative formalism.

The foregoing critical statement demonstrates the connection between language and physical movement, while also suggesting questions not specific to Dante alone: how does language mean and what can language actually describe? The quest narrative, a common motif in allegorical narratives, literalizes this tension between language and movement. The most common varieties of quest narratives involve journeys through landscapes which, if foreign or unfamiliar, contain tangible details. Even Spenser's most fabulous creatures terrorize recognizably English landscapes. Likewise, while early modern travelers may have described cities populated with cannibals and human-animal hybrids, they evoked the details of these new places in comparison to Europe. In the preceding chapter, I began to explain the energetic dynamics of movement. I will here fill out the thematic parameters of such movement in allegorical narratives that take their protagonists beyond earth.

The cosmological allegorical narrative is, like many allegories, a form of the quest narrative. The medieval variant takes the form of a quest through "the totality of [. . .] 'natures' [. . .] arranged to form a hierarchically ordered whole or cosmos" postulated by Aristotle.[6] However, the concern inherent to this narrative model with the limits of representability and representation distinguishes it from other quest narratives. In this type of narrative, the actual subject depicted and the language used to depict the subject are called into question. In the standard quest narrative, which perhaps we imagine taking place in some *selva oscura* filled with dense oaks and foreboding elms, a firm correlation exists between the dark (but nameable, identifiable) trees of this forest as signifieds and signifiers. I employ this example of the trees in the *selva oscura* to evoke Saussure's own tree diagram in his description of the nature of the linguistic sign in the *Course in General Linguistics.* Perhaps, in keeping with Louis G. Kelly's "Saint Augustine and Saussurean Linguistics," we should regard this tree as the tree associated with the Fall.

In the quest narrative, language itself, with its limits, does not serve as an object. The referentiality of the Grail, for instance, is not a significant motivating factor of the Grail quest. The Grail knights do not undertake this quest with the intent of exposing the Grail as a fraud, revealing "Grail" as another word for cup. Even Indiana Jones' Grail quest demonstrates that the Grail, housed in the sumptuous but barren desert palace of Petra, is a relic with formidable powers. In such a narrative, the object of the quest, regardless of its mystical powers, serves as a stable linguistic and wholly textual sign. Tzvetan Todorov argues that "the quest of what the Grail means is never over, and we are continually obliged to relate this concept with others which appear in the course of the text."[7] For Todorov, the Grail operates as a signifier which signifies and drives narrative. The Grail as concept is never completely revealed; however, this quality of concealed meaning produces the quest narrative. In the cosmological allegory, such a clear pattern of signification does not exist. Indeed, the pairing of concept and sound-image which constitutes the linguistic sign also constitutes an arduous process which we could define as one of the key characteristics of the cosmological narrative. In this type of narrative, concept and sound-image elude one another. Saussure's concept tree, for example, besides corresponding to the sound-image "tree," also implies the terrestrial qualities of the tree such as rootedness and photosynthesis. This diagram reinscribes the terrestrial presuppositions of signification.

The protagonist of the cosmological allegory, on the other hand, encounters a celestial body that, unlike the page, remains luminous and unmarked while producing meaning. Italo Calvino expresses this paradox in "The Tale of Astolpho on the Moon." In this story from *The Castle of Crossed Destinies,* the narrator evokes the moon as the origin of myth and narrative, remarking that "From this arid sphere every discourse and every poem sets forth; and every journey through forests, battles, treasures, banquets, bedchambers, brings us back here, to the center of an empty horizon."[8]

The celestial voyager in the medieval cosmological narrative navigates an area where linguistics and cosmology directly intersect. Such a narrative expresses through a language based on referentiality that which is completely beyond lived experience. This distinction is of key importance for our attempt to consider these cosmological narratives as allegories. Indeed, critics define traditional personification allegory by a stable relationship between referents and meanings. Thus, the interpretive scaffolding of the traditional allegory relies on a determinable hierarchy of referentiality. This relationship must be stable in order to convey the intended scriptural vision. As Rosemond Tuve notes in *Allegorical Imagery:*

> Most images that undeniably require or allow allegorical reading convey in this manner, through public figures or symbols, the needed concepts which trigger thought. The thoughts are not only judgments or evaluations on moral matters, but are ideas concerning ultimate destiny, divine beings, supernatural forces. These seem to need a surer and richer language than any private [. . .] symbolism, since the interpretation of the literal with the metaphysical, as well as the moral, meanings is an element we persistently find in mediaeval allegory.[9]

Critics perceive modern or postmodern allegory as foregoing this stable hierarchy. The decidedly allegorical images of an artist such as Remedios Varo, for example, rely on an interpretive and creative process that places traditional symbols and tropes within a highly personal, but still recognizably allegorical, interpretive framework. Such a strategy results in new relationships between images, allowing for a process of allegorical resignification.

This narratological concept of the resignification of an allegorical motif also informs our understanding of the cosmological dream narrative. However, in this narrative, the writer produces more than a map of an individual psychology. Above all, the allegorist employs the mode to examine how human language represents the divine. The act of navigation itself intersects the linguistic and the cosmological. The process of signification in these narratives always moves toward the impossibility of its own success. As the voyager moves farther away from waking reality, entering a dream reality which enables the celestial journey, he becomes unable to map a realm which gives order to that waking reality. He is inspired by, but unable to totalize, the domain of the *oneiros.* Thus, in this type of allegory, both the character and the language used to depict that character are engaged on a journey which can never be conclusive, can never lay hands on, let alone drink from, the Grail. Paul Ricoeur emphasizes the extent to which this dominates dream analysis, writing in "Energetics and Hermeneutics in *The Interpretation of Dreams,*" "Analytic interpretation must be supplemented by a genetic interpretation [because] symbols have a special overdetermination which is not the product of the dream-work but a pregiven fact of culture: they are often the vestige of a conceptual and linguistic identity now lost."[10] In the dream narrative, the voyager describes while moving steadily toward the indescribable. Likewise, the allegorical journey to the stars begins with recognition of the limits of language and the unimaginable shape of a divine cosmos.

In pairing these, the narrative attempts to render language complete and the cosmos knowable. To this extent, the map of the stars is a fundamentally linguistic model. For the medieval writer, language itself was already stellified. Categories of philosophical concepts, themselves shifting and multivalent when applied strictly to terrestrial phenomena, became clearly delimited when mapped onto the universe, and conformed to shapes predetermined by the divine. In the cosmological narrative, words transmorph into star charts. The celestial dream journey, then, necessitates an interpretive strategy which differs from that of more terrestrial quest narratives.

In the following sections, I will begin with a discussion of Augustinian sign theory and then move to the narrative devices used to propel allegorical characters into the stars. Augustine's importance, summarized by Stephen Prickett as follows, attests to the influence of sign theory on states of consciousness: "An interiorized self-consciousness goes back at least as far as St. Augustine's *Confessions* [. . .] Augustine's most revolutionary contribution to human thought lay in the discovery—or rediscovery—of introspection."[11] I assert that the allegorical motif of celestial travel embodies the cognitive experience of introspection and the singularity of consciousness impelled by classical and Christian sources.

THE TRANSCENDENTAL SIGN

Augustine's interest in linguistics and time helps to clarify this dilemma. First, Augustine's definition of the sign reflects a common medieval view of the relation between linguistic expression and material reality. Indeed, the Neoplatonists modeled their ideas regarding these matters on Augustine's works:

> Augustinian theology, echoed in Christianized natural science, shares with Platonic and Neoplatonic Pythagoreanism the fundamental mode of thought which differentiates the medieval from the modern imagination, a mode of thought which is likely to strike us, the heirs of Galileo and Newton, as disturbingly 'unscientific.'[12]

In *On Christian Doctrine,* Augustine defines the sign as "a thing which causes us to think of something beyond the impression that the thing itself makes upon the senses."[13] In this definition, Augustine's reference to the "impression" made by the sign is not without consequence. Meanings embed themselves and exert a force on the individual; likewise, "signs, which belong to the realm of the body, partook of the ambiguity and opacity of things as they appeared to the fallen bodily senses."[14] Further, Augustine's reference to the sign as something beyond the thing it signifies recalls the fundamentally allegorical consciousness animating signification. At the same time, Augustine's presentation of the devil as rhetorician (see, for example, *On Free Will*) coincides with a projection of "the crucial role of verbal persuasion in the Fall."[15] Meaning, derived through interpretation, testifies to a divine elegance unbound by the materiality of the sign.

As Eugene Vance notes in *Mervelous Signals,* "Augustine's tendency to treat things not for themselves but as signs remained a habit of medieval culture during the centuries that followed."[16] Furthermore, Vance notes that "Augustine considered verbal signifiers, *voces,* to be corporeal things, even though what they signify is not corporeal but mental."[17] This intrinsic relationship between signifier and signified helps explain the limits of the cosmological journey. Both the places beyond the celestial sphere and the words used to describe them lie beyond the bounds of representation. Likewise, this boundary, unseeable when considered from a purely linguistic perspective, itself becomes a real thing when placed within the frame of the cosmological model. The limits of the language and figuration of cosmological or cosmographical narrative throws in relief an equally rarefied, presumably literal envelope—the celestial, astrophysical envelope surrounding Dante's actual cosmos.

Thus, we can view verbal signifiers as corporeal things in the Neoplatonist conception of the universe. The boundaries between categories of signifiers were also linked to very real components of the physical universe. The injunction against what was representable and what was not depicts a sense of harmony in the shape of the universe that likewise extends to language. We could view the obvious limits of language as a sign of its failure. However, for medieval commentators, such limits and solid boundaries hint at the essential harmony of the universe. A fallen humanity should not be able to impress the image of the divine within language. Language itself, as a gift of God, can conceal mysteries that are spatially concealed beyond the celestial sphere. Thus, "all harmonies, within the soul or without, are only manifestations of a universal harmony called reason (*ratio*) and this harmony transcends all others."[18] The harmony of the universe is essentially linguistic in its composition as well. In Robert Jordan's words regarding the Christian influence on the cosmic symbolism of language:

> Though belief in the divinity of the Bible was not an element of the Pythagorean heritage, the Christian mode of implementing this belief was consistent with the symbolic rationalism of the Pythagoreans. Indeed, the symbolic interpretation of language is no less venerable than the symbolism of number. But antique studies of rhetoric did not give rise to a cosmic symbolism comparable to Pythagorean mathematics and number theory, the reason being inherent in the respective possibilities of number and language as images of perfection. Nevertheless, belief in the symbolic value of revered texts was part of the classical heritage of the Christian Middle Ages. Like number theory, textual symbolism began with the assumption that the text as perceptible object mirrored a higher intelligible truth.[19]

Allegory thus serves as the semiotic system par excellence. For Augustine, humankind uses language in an attempt to reconcile sinful mortality and the Divine.[20] Indeed, Augustine's writings on dream signification accentuate this position. As Martine Dulaey points out in *Le Rêve dans la vie et la pensée de Saint Augustin*,

> Augustin avait réfléchi sur le rêve dans la Bible. Toutefois, son étude n'a rien de systématique. Il ne cite qu'un petit nombre de textes bibliques concernant le rêve, toujours les mêmes, qu'il utilise de façon identique: pour illustrer l'opposition des visions spirituelle et intellectuelle.
>
> Augustine has thought about the dream in the Bible. However, his study is not systematic. He only cites a small number of Biblical texts concerning the dream, always the same, which he uses in an identical manner: to illustrate the opposition of spiritual visions and the intellect.[21]

However, while "man labors [. . .] in a poem of history that he cannot read as a whole,"[22] the spiritual realm offers the promise of a language that reveals all without concealment. In this realm, "there is no difference and [spiritual beings] understand God's logos as a discourse proffered without syllables, without syntax, and without enigma."[23] Paradoxically, this lack of distinction and differentiation makes the language of the celestial sphere so unknowable. It signifies in a single and absolute moment, while humanity moves through time using a language built from oppositions and partial meanings.

This characteristic of Divine language still animates the manifestations of reason in a fallen world. In his *De musica libri sex,* Augustine links the shape of the universe to this harmony of reason:

> From there come all times which are formed [*fabricantur*], ordered, and regulated [*modificantur*], like an imitation of eternity, as the revolution of the sky returns to the same point and brings back to the same point the celestial bodies, obeying, by means of days, months, years, and lights, and other astral movements, the laws of equivalence, unity and order. Thus, the things of the earth are subjected to those of the heavens, and by the harmonious succession of their times they associate their movements with a kind of poem of the universe.[24]

This statement also emphasizes the role of time or movement in language and representability. Time, itself caused by the movements of the planets, echoes a divine harmony beyond time.[25] Thus, while "creation exists in time," an "indivisible present" encircles time, revealing its finitude.[26] Further, in this statement, Augustine emphasizes the mechanical or preconstructed nature of the universe. Time itself has been fabricated by the laws of astral movements.

But laws, as medieval commentators would be well aware, are based on prohibitions and oppositions. A divine order without syntax, syllables, or enigma differs significantly from a natural order built quite mechanically from the very intersection of these elements. At the beginning of this chapter, I made a distinction between the medieval cosmological journey and the more modern incarnation of the same allegorical motif. The distinction I made between the two was based on the fashioning of machinery to ascend to the stars. Medieval cosmological allegories, while not machineless, rely primarily on the dream-space as the method of astral propulsion. In modern science fiction, on the other hand, fully conscious subjects tinker away on machines, unaided by personifications of Truth or Justice or the Seven Liberal Arts. These representations of movement, the narrative motifs used to transport characters, reveal fundamental philosophical differences between these genres.

I'd like to explore this distinction a bit further as we have just considered the mechanical order of the universe and the limits of language. In his sign theory, Augustine highlights written language. The concept that signs are themselves things corresponds with the well-documented textual focus of Christian philosophy. The written text, furthermore, enjoyed a legitimacy not accorded to most speech acts. Even the Word of God only gains absolute legitimacy when made flesh through the Incarnation. Jacques Derrida, describing this logocentrism in *Of Grammatology*, argues that "the age of the sign is essentially theological."[27] Indeed, the Word of God is written on Christ the lamb in the same way that Biblical manuscripts were carved into vellum, the processed and treated skins of sheep. But where does the dream fit in as a speech act, or even as a technology? Even Macrobius' authoritative discussion of dreams in the *Commentary on the Dream of Scipio* reveals two different sets of standards for analyzing the dream as manifestation of the unconscious and dream as rhetorical strategy or narrative mode. Still, the dream as a preferred mode for communicating the shape of divinity or receiving the communications of a divinity beyond the limits of speech obeys the same rules and limitations as the (super)natural order it expresses.

THE IDEAL SIGN OF DIVINITY

Invectives made against the *narratio fabulosa* as a vehicle for describing the divine emphasize the blasphemous nature of transcending the transcendent. But poetic language allows the philosopher to pursue this goal while masking his/her intention in allegory's iridescent robes. Macrobius suggests that when writers "wish to assign attributes to these divinities that not only pass the bounds of speech but those of human comprehension as well, they resort to similes and analogies."[28] But not all similes and analogies are equal in their ability to represent the unrepresentable. Plato, Macrobius tells us, found only one image suitable for an accurate depiction of this highest cause: "In truth, of visible objects he found the sun most like it, and by using this as an illustration opened a way for his discourse to approach what was otherwise incomprehensible."[29] Macrobius' reference here corresponds to Neoplatonist concerns.[30] As Stahl notes, "it is a common practice of Neoplatonists to foist upon Plato philosophical tenets which originated in their own school."[31] Still, the image that Macrobius provides for us serves to clarify the link between astronomy and semiology. The divinity of the sun is not, of course, a new idea, or even one specific or at all new to the Neoplatonists. The significance of the image stems from the peculiar hybrid of hermeneutical analysis and interest in the natural world espoused by early Christian commentators, such that "the Sunne, filled with a divine vertue, wonderfully imparteth the same unto these inferior bodies."

Neoplatonists used Macrobius' *Commentary on the Dream of Scipio* as an authoritative work on an encyclopedic variety of subjects. Likewise, it provides a bridge between pagan sources and Christian interests. Stahl appears critical of this appropriation and transformation of concepts, but this process was central to the continued relevance of classical works. Macrobius' selection of the sun as the ideal motif for the Divine speaks to the connection between medieval sign-theory and the cosmological dream allegory. The sun is, first, a sign for depicting Divinity and second, a celestial body.[32] This order of significance clarifies the interests of early Christian thinkers. As Rita Copeland indicates in *Rhetoric, Hermeneutics, and Translation in the Middle Ages,* "early Christian semiology accords human language a secondary, although necessary, role in relation to the primacy and stability of divine signification."[33] The divine aspect of signification endows certain signifiers with a stability missing from others which have no clear theological purpose.

The sun then signifies primarily, and most importantly, as a signifier equated with divinity, its qualities equally illuminating and blinding, a synthesis of the *via positiva* and *via negativa.* The anonymous *The Cloud of Unknowing* (1370) employs a qualititatively different image to explore such a synthesis. Here, "A Beam of Ghostly Light" (87) pierces the dense cloud of mortal life. The sun's identity as a ghostly or spiritual body necessarily entails a recognition of its primary meaning as a marker of divine signification now in a Christian cosmology. Further, the stability of the sun as a component of the natural world, itself fallible and subject to death and decay, conveys its ultimate stability from a theological standpoint.[34] In other words, for medieval thinkers to refer to the sun and other celestial bodies as parts of a natural system, they had to first consider the significance of these bodies as theological concepts.

However, even the intricate theory of celestial correspondence could be revised for further theological illumination:

> Surprisingly enough, no author prior to Dante seems to have made what would otherwise appear an irresistible symbolic and theological concordance: the assignment of each order of the angelic hierarchy to one of the celestial spheres, matching them by their relative position within the order of spheres or the angelic hierarchy.[35]

This navigation between seemingly separate systems of signification, discussed in medieval semiological theory, drives medieval allegorization. To this extent, "twelfth-century allegorization [. . .] is frequently distinguished less by abstract transfers or transparencies than by subtle tensions of thought and nuance."[36] Whitman takes the viewpoint that this process does more than attempt to verify concepts through appeal to the *auctores*. Instead, medieval semiologists and allegorists expressed older concepts in new ways made possible by Christian theology:

> When writers such as William of Conches and Abelard interpret an ancient figure like Plato's 'World Soul' as the 'Holy Spirit' or divine 'love,' for example, their new composite points ambiguously in more than one direction, both toward the worldly animation of the pagan figure and toward the otherworldly inspiration of the Christian one.[37]

Whitman suggests that this division between the pagan and Christian significations of a figure further complicates the expression of concepts in medieval allegories. He provides various examples which could be attached to the sun as an example of divine signification. However, the distinction between the World Soul and Holy Spirit leads to a fusion of meanings which complicate the relation of the signifier to the divine. Thus, the signifier which represents both the World Soul and the Holy Spirit, two transcendent concepts, is, from a purely Christian perspective, torn between the worldly and otherworldly.

The shape of the cosmos proposed by medieval allegorists reflects this largely linguistic problem. As Whitman explains, "the multivalent figures in such allegorical interpretation imply a multilayered cosmos."[38] The shape of the cosmos corresponds with the polysemy of linguistic meaning: the two are built from one another. But the complexity of the interaction between these two stems from the attempt by Christian writers to reconcile a nascent

Christian philosophy with classical sources. The Christian sign system, ostensibly based on the formulations of the *auctores,* also reveals fundamental differences regarding the valuation of both concepts and phenomenon. Further, the medieval reliance on the rhetorical patterns of *amplificatio* and *descriptio* does little to reconcile these opposing meanings. Instead, such meticulous catalogues expose diverse discursive approaches as a way of exploring these competing theories of being.[39]

Cosmographers of the twelfth century were quite interested in resolving two competing systems for describing psychological states and physical phenomena. Thus, the logical structure of Platonism provided an analytical structure for describing the mechanics of language and the mechanics of the cosmos. However, the actual concepts associated with the logical structure of Platonism were not quite in accordance with the Christian attitude toward the Word and the interconnection of language and spirit so central to Christian conceptions of creation and the natural world. The Neoplatonic concepts for explaining the structure of the universe did provide a rhetorical model for expressing such concepts.

In other words, the Neoplatonic "map" of the universe, because of its emphasis on correspondence between the various levels of the microcosmos and the macrocosmos, influenced the rhetorical modes of discourse available to Neoplatonist writers. This correspondence between (1) the physical shape of the universe, (2) humanity and the cosmos, and (3) human language as a direct expression of such correspondence stemmed from "the ancient Neoplatonic notion that each distinct level of 'procession' and 'return' in the cosmos has its own logos—both its own articulation and its own rationale."[40] Thus, the movements of celestial bodies are linked to a logic which parallels the logic of representation. Twelfth-century Neoplatonists tended to treat "the universe as an intricate system of divine unfolding (*explicatio*) and folding (*complicatio*), in which the different modes (*modi*) of the divine order correspond to the different modes (*modi*) of human comprehension."[41]

This suggests that the shape of the universe expresses specific discursive modes. Divine order, posited as the fundamental governing principle of the universe, organizes matter in a manner that parallels language itself. Thus, the workings of the universe and the well-documented varieties of discourse as found in the many handbooks on rhetoric at the time are described as modes of knowledge.[42] The connection between these two suggests that the universe operates like language and that the modes of discourse outlined by rhetoric are themselves natural and fundamental components of human consciousness. Indeed, the similarities between the universe and discourse stem from a dichotomy perceived as belonging to each. While the cosmos and language stem from a single unknowable source, they present an exterior surface, or veil, which could be examined and could, perhaps, give way to this mystery.[43] Whitman stresses this correspondence between the shape, structure, and movement of celestial bodies and the shape, structure, and delivery of human language.

This abstract correspondence between cosmology and semiology informs the examination of allegorical texts set beyond naturalistic or "realistic" landscapes. The types of rhetorical approaches that predominate in these texts hint at the high degree of interconnectedness between the shape of the universe and the form of its representation as text. The universe was not merely presented as a system of divine folding and unfolding. Instead, such ideas likewise animate the language used to describe such processes. Whitman, following the lead of twelfth century cosmographers, describes the intricate system of the universe with rhetorical terms. Thus, the universe becomes a system constituted through a dialectical relationship between microcosmos and macrocosmos as well as *explicatio* and *complicatio.*

Microcosmos and macrocosmos are familiar terms: they indicate the interaction of levels of celestial correspondence. Even when referring to correspondence between the human soul and the shape of the universe, they suggest a system of binaries that does not necessarily conceal supernatural meaning. Indeed, the development of artistic perspective is clearly linked to this correspondence between the natural and the human. Further, early works on perspective indicate that these correspondences can be known and explored to arrive at the expression of order and symmetry. *Explicatio* and *complicatio,* however, do not suggest such direct correspondence. Instead, the dialectical relationship between these two types of communication suggests hidden and revealed meaning. Further, their use as ways of describing the physical world suggests that the universe itself is a primarily textual language act constituted by this complex relationship between revealed and concealed layers of meaning.

The relationship between *explicatio* and *complicatio,* like that between microcosmos and macrocosmos, suggests interpretive movement. The implementation of *explicatio* and *complicatio* to describe the physical world stems from the correspondence of the phenomenological and textual universe. The effectiveness of an evocation of microcosmos or macrocosmos comes only through a comparison to the opposite item. Thus, the dimensions of a building, a cathedral for example, grow in significance through a demonstration that they correspond to the dimensions of the macrocosmos. The movement between microcosmos and macrocosmos constitutes this relationship: there is no microcosmos without macrocosmos and vice versa. Such a relationship positions the cosmos itself as an intermediate indeterminate process which is this movement between two opposing but complementary levels of representation of that process. The cosmos exists, then, as a series of interpretive movements between two opposing but complementary levels of representation. Both macrocosmos and microcosmos refer to a process of representation that is always in flux because it is intended to invoke an opposing end of a spectrum of scale.

As rhetorical methods, *explicatio* and *complicatio* reflect one another. The most cogent *explicatio* reveals further dimensions of complexities. In this sense, *explicatio,* while described as a means of revealing meaning, is not a process which produces textual or hermeneutical closure. Likewise, consideration of language and universe as parallel suggest that each is a complete and perfect system. As Whitman points out, for twelfth century cosmologists, *explicatio* as a mode suggests movement and exploration. Invested in the creation of texts that can be classified as imaginative literature, philosophical investigation, and scientific observation, twelfth century cosmologists "treat the act of explicating ancient texts, 'uncovering' their integumenta, as virtually coextensive with the act of discovering the natural world itself."[44] The grammar of the natural world, as revealed by these cosmologists, remains a perfect grammar, an ideal text.

The process of examining the natural world itself changes precisely because of the system of hermeneutical analysis applied to natural phenomena. The mix of imaginative narrative, theological doctrine, and philosophical inquiry which constitutes the method of twelfth century cosmologists forces a reconsideration of the textual status of the natural world. From a purely Biblical standpoint, the natural world is a finished text written by ineffable Divinity. Thus, the text itself, the natural world, is readable, rendered legible through sensory perception. However, the grammatical system used to write that text remains beyond the limits of the knowable.

Or that would appear to be the case. Instead, the very interpretive apparatus appropriated for the examination of nature as text reveals instead nature as system and not text. The move from nature as specific text to nature as grammatical system implies a linear development from the hermeneutical approach to natural phenomena. This development stems, in part, from one of the dominant purposes of the Neoplatonists: the recovery of classical sources within a Christian philosophy.

The pairing of classical and Biblical figures, so prominent in Renaissance art, suggests the range of this intellectual preoccupation. Erich Auerbach's *Mimesis* provides perhaps the most well-known analysis of the underlying rhetorical and structural components which differentiate Biblical and classical narrative. For the purpose of this discussion, we must also consider the perception of the creation of nature in classical myth and Biblical history. From a purely rhetorical standpoint, we might comment on the dominance of *prosopoeia* in the accounts of the creation in classical myth. Further, not only are natural processes embodied in recognizably human forms, they are embodied in recognizably gendered forms. Thus, father sky and mother earth produce the natural world through their coitions. In the Biblical story of creation, we are left with a much less clear anthropomorphism. While humans are created in the image of Divinity, that image itself remains unclear. An indistinct spirit form merely moves over the waters, giving rise to nature. Within the Christian context, the creation of the natural remains an act bound within a historical continuum. The creation occupies a definite point on a finite timeline. The more metaphorical process of creation provided by classical myth suggests a process that is continuous. The sexual motif closely parallels a seasonal death and rebirth that, while indicative of spiritual rebirth in a Christian context, implies, in classical myth, the continuous creation and recreation of the natural world.

The model of the cosmos as a text so frequently employed in cosmological dream allegories results in a crucial paradox between ways of representing the interpretability of that text. The combination of genres and approaches characteristic of twelfth century cosmologists results in a model of the world as a text perpetually in the midst of revision, and this despite the inherent perfection of that text, established by a purely Biblical conception of the universe as a monolithic text, engraved and uneditable but still mysterious and ineffable. While medieval commentators would certainly be aware of the extent to which the Bible as text is itself a construct cobbled together from numerous sources and over many centuries, they would also acknowledge that the finished text provides the means to approaching theological truth. The combination of sources and approaches guiding these cosmologists, however, results in a model of the universe that, like a text, is continuously 'texit et retexit,' or 'woven and unwoven.'

The competing viewpoints of the universe as knowable text and ineffable mystery appear together in the works of twelfth century cosmologists. We note the appearance of these viewpoints through two dominant motifs. The first motif is anthropomorphism itself. This pairing of classical myth and Biblical history led medieval scholars to favor classical sources and approaches to subjects to, as Roger Parr argues, the point of slavish imitation.[45] The anthropomorphism of the natural provides a form that, through allegory, approaches the philosophical complexities of classical texts. Thus, "with the vivid allegories of Bernard Silvestris and Alain of Lille in the middle and later decades of the twelfth century, the process of philosophic explication dramatically converges with the process of imaginative composition."[46] The second image, of the world as book, lies between two competing means of interpreting that book. The rhetorical methods of *explicatio* and *complicatio,* for example, were used by cosmologists to explain the universe in literary terms. This impulse is most familiar through the representation of the world as a rhetorical performance.

The anthropomorphism of the components of the physical world in cosmological allegories reinstates the significance of this motif. Textual practices directly influenced a perception of the limits of the natural world. In Alain de Lille's *Complaint of Nature,* for example, the natural world, as the dominant speaker, descries the "blindness of ignorance, [. . .] delirium of mind, [. . .] and impairment of sense" afflicting humanity.[47] The multiplicity of tropes used by Nature, the consummate rhetorician, testifies to the complexities of the natural world and the vast catalogue of rhetorical strategies which twelfth century writers were most eager to employ in their works. Catalogues of motifs and rhetorical and oratorical strategies were as central to late medieval conceptions of language and the natural world as theological texts which acknowledged the limitations of language in describing and explaining the natural world.

But there is an incongruity between an anthropomorphized nature able to orate at levels which even Cicero would admire and a Divinity, represented by the sun, which is both speechless and beyond speech. Both motifs are present in cosmological dream allegories, and indicative of two poles of communicative possibility enabled by the physical world, either a personified figure that communicates like a text, or a form that cannot be expressed in a text. And, in fact, this characteristic of the genre comes to the foreground in Kepler's *Somnium.* Most specifically, the figure of the Daemon seems to straddle these two poles of possibility. As an agent of the physical world, the Daemon communicates like a text; we cannot fully separate its speech, a scientific discourse on lunar astronomy, and its form, an unseen essence that could be physical or spiritual. To this extent, the questions underlying representation that I have explored in this chapter remain as pertinent for Kepler as for his predecessors.

Chapter Four

The Process of Stellification

STELLIFICATION AND THE DREAM BOOK

After a long series of invocations and introductions, the narrator of Chaucer's *The House of Fame*[1] finally begins in Book II "to telle al my drem aright."[2] The dream continues as a giant golden eagle, which recalls Dante's Eagle in *Purgatorio,* swoops down and steals the narrator in "hys grymme pawes stronge."[3] At this point, the narrator is clearly afraid of many things: the eagle may drop him, he may be forced into service, like Ganymede, as "the goddys botiller,"[4] or he may be transformed into a star or constellation. Chaucer ponders his fate in relation to the significatory value of his journey. He is curious to know "wher Joves wol me stellyfye or what thing may this sygnifye."[5]

The word "sygnifye" follows "stellifye" ostensibly in keeping with the poetic form of octosyllabic couplets that characterizes *The House of Fame.* At the same time, both words refer to meaning, to the relationship between a thing and a word-image. This definition is clearly recognizable for "sygnifye," but the mythical process of stellification presupposes a similar discursive process. While the narrator refers specifically to Ganymede, many mythical figures met stelliform fates. A list, by no means comprehensive or complete, includes Orion, Aesculapius, Andromeda, Callisto, Arcas, Castor and Pollux, Romulus, and Cassiopeia. Ovid likewise evokes Caesar's soul, "snatch[ed] up [. . .] from his murdered body, and transform[ed] [. . .] into a star."[6] Along with bodies transformed into stars, classical mythology also includes examples of the astral transformation of objects closely related to earth-bound characters. Thus, both Ariadne's crown and Orpheus' lyre gleam in the night sky.[7] These heroes, and their accouterments, shine in glory among the heavens and as names evoking myths attached to physical things.

For Chaucer, the threat of stellification proves unfounded, as the Eagle soon reveals that "Jove, thorgh hys grace, wol that I bere the to a place which that hight the House of Fame."[8] Notice that Chaucer uses the verb "hight" to designate the House of Fame. This word choice points to a place name. But, in conjunction with "sygnifye" and "stellifye," the word implies a different kind of cognitive process. The House of Fame, although an allegorical and imaginary setting, can be named without necessarily meaning anything beyond its name: it is both an allegorical signifier and a marker of enclosed space.[9]

In contrast, the identification of a mythic character with a star always begins a process of allegorical signification. The sphere of the heavens is an actual place, its movements impacting individual destinies and weather systems. At the same time, the characters stellified by myth, these earthly inhabitants of the heavenly spheres, transform this place through their own transformations. The identification of a character, and the story attending his/her metamorphosis, results in an expansion of the myriad narrative possibilities in the patterns formed by the stars.

While I, following Chaucer's lead, have referred thus far to the transformation of mythical figures, the mission of the protagonists of medieval cosmological allegories differs considerably from the after-death astral enshrinement offered by the classical gods.[10] In this genre, popular until the seventeenth century, the narrator actually travels to the stars, bypassing the blinking beacons of fallen heroes along the way. While Orpheus' lyre performs the music of the spheres nightly, Orpheus himself never ventured so far. Even his journey to the underworld, a supernatural space like the heavens, remains earthbound. Greco-Roman mythology, rooted to the Mediterranean, recognized Avernus, the entrance to the infernal regions, as "a miasmatic lake close to the promontory between Cumae and Puteoli, filling the crater of an extinct volcano."[11]

In this genre, then, space exists and fills an important role as an attainable supernatural setting. This is not to say that there are no narratives of astral travel until the medieval era, however. One example in particular, Plutarch's *The Face in the Moon*, significantly informs Kepler's *Somnium*. The pathway to the stars reveals the psychological distinction between the medieval and modern perception of this starscape. The medieval and early modern protagonists of cosmological allegories have recourse to two basic methods of astral propulsion: the monstrous machine and the dream. At first glance, these appear very different: one is active, transporting its cargo and flapping its wings, while the other is passive; one is supernatural, rarely encountered except by visionaries and heroes, while the other is common, encountered nightly; one places the fantastic in the real while the other eludes the real through the subconscious machinery of the dream. However, the similarities, and also the interpenetration of these categories, serve to distinguish the cosmological allegory, regardless of the means of conveyance, from the narratives of nineteenth and twentieth century cosmonauts.

Alain de Lille's *Anticlaudianus*, for example, provides the clearest example of this distinction, a distinction not specific to *Anticlaudianus*, but also appearing in, among others, the *Hyperotomachia* of Francesco Colonnas. Describing these two texts, Peter Dronke notes "The 'hermeneutic' writers [. . .] had the impulse to create a language mosaic, in which archaisms, coinages, graecisms and glossary-words were set as so many exotically coloured stones, a language also where syntax tended to become grandiose, flamboyant and at times impenetrable."[12] Dronke's use of the phrase "language mosaic" suggests the extent to which the stellar journey and the representation of that journey coincide. Words are "set" in a text in the same way that the stars are "set" in the unmoving outer sphere. Dronke's description of twelfth century poetics highlights the imaginary components of such narratives. H.R. Jauss likewise notes that in distinguishing the fictive from the actual, the Chartres school enabled the imagination, granting it "the status of a creative capacity belonging to the poet, who is able to illustrate ideas and archetypes (those with which Neoplatonic thought had populated the Christian cosmos) through his use of the figures and fictions of rhetoric and, finally, through the fictive discourse of a *fabula*,"[13] a process clearly exemplified by Alain's *Anticlaudianus*.

In the *Anticlaudianus*, the Seven Liberal Arts construct a chariot built according to the rigid specifications of reason:

> As rule requires, order demands, reason requests, Prudence orders, the aforementioned band of sisters, working together with finer file, moulds and remoulds, fashions and refashions, decks and bedecks the parts of the chariot. They remove from them every appearance of irregularity of form and every flaw that would invite criticism.[14]

Alain emphasizes the physical form at the center of the cosmic vision, supplanting the vision itself as an enabling feature of the astral ascent. Instead, Alain stresses the physical form of the chariot, providing details of the "articulations, nails, and connections [. . . which] unite [. . .] the scattered parts in order, bind [. . .] them by law, and fit [. . .] them in place."[15] The classical ideal of artistic wholeness applies to the dimensions of a chariot which is eerily supernatural in its perfection.

Like the computing engine of Ramon Lull, the chariot takes the form of "a cybernetic machine, prepared to unravel every problem, every science, even faith itself."[16] As Ambroise Paré defines the monstrous in *Of Monsters and Prodigies*, perfection and order "doth abound" in this chariot: Nature herself has no part in its construction. At the same time, the classical ideal of artistic wholeness applies to the dimensions of a chariot which is *complete*: as an exemplum of knowledge, the chariot is a text which includes every conceivable combination of character, number, word, and figure demonstrable by the Seven Liberal Arts. This vehicle allows Alain, through the character of Phronesis, to pass through and describe the planetary spheres. However, once Phronesis enters the celestial sphere, she allows Theology to guide her to heaven. Alain here abandons the chariot, an assemblage indebted to the various levels of human knowledge, to a force beyond the rational.

Alain's account, like many others, features a vehicle that, on one hand, anticipates the devices employed by contemporary figures, ranging from Jules Verne to Wallace and Grommit (who travel to the moon for a cheese holiday). Indeed, the flying machines of narratives such as the *Anticlaudianus* or Ariosto's *Orlando Furioso* prefigure the technical machinery appearing in the notebooks of Leonardo da Vinci. At the same time, narratives such as Chaucer's *The House of Fame* and Dante's *Purgatorio* feature flying animals as the preeminent means of astral conveyance. If we regard the chariot only as machine, we would distinguish it, as motif, from the great eagles or talking daemons that likewise occupy this role of cosmic vehicle. The Gulden Draak, a huge metal dragon perched atop the Belfort Tower in Ghent since 1382, perhaps suggests this fusion of supernaturally monstrous animal and machine that I am evoking here in order to nuance our understanding of the astral vehicle as motif. Such creatures play a key role in the medieval imagination, existing at the permeable boundary between the natural and the supernatural. For Timothy Beal, author of *Religion and Its Monsters*, this boundary is itself fashioned from the skinned carcasses of monstrous bodies. Indeed, the complex structures which constitute culture and the realm of the civilized are themselves engaged in an ongoing process of "emerging out of chaos."[17]

Alain explicitly conjoins the machine-like chariot and the human body, viewing the body, as Silvestris concurs in the *Cosmographia*, as "the masterwork of powerful Nature."[18] At the same time, while body and machine exist as masterworks, they only achieve their full potential through the abandonment of the rational natural order through which they are constructed. In this way, the chariot of Alain, an object at the center of a visionary narrative, calls attention to the second, seemingly less technical, mode of transport available to the medieval cosmonaut: the dream.

The process of stellification, and its implications for our reading of the textual features of imagined astronomical journeys, occurs regardless of the mode of transport. Indeed, the relation between the two is actually quite close, as evidenced by the word "stellify" itself. While etymologically related to the Old French *stellifier* and medieval Latin *stellificare*, the verb *stellify* makes its first English appearance in Chaucer's *The House of Fame*. Chaucer, building from mythical sources, employs the word in a strictly poetic sense. For Chaucer, as well as later writers such as John Lydgate (in *Temple of Glas*, c.1403) and George Cavendish (1562), the word recalls direct Ovidian metamorphosis into a star or constellation.

In the seventeenth century, the word also develops a connotation more directly connected with the stars as celestial bodies and not mythic figures, referring to, as in T. Blount's *Estienne's Art Devices* (1650) a "Devise wherin was depainted the Skie stellified." In 1562, G. Cavendish still employs the word as Chaucer does in 1384. In his *Poems*, he describes a "lady most excellent, by vertue stellefied, Assendyng the hevyns, where thou raynest aye." This lady, like Beatrice, like any of the mythical figures astrally transfigured, assumes a position as an allegorical personage, her virtue not merely personified, but also stellified. In 1658, on the other hand, Phillips uses the word in the service of astronomy: "That without stars is the Primum mobile, the other are all stellified, either with fixed stars or planets." Clearly informed by Neoplatonism, the astronomy here refers to fixed stars and the Primum mobile. This is not to suggest, however, that the word and its derivitives become immediately incorporated into the language of the *nova astronomia*, gradually supplanting Neoplatonist cosmology while using language indebted to its philosophical traditions. In fact, stellify continues to be used poetically, so that, in 1650, J. Reynolds refers to the "rejoycing of joyful subjects," however redundantly, as an act of the "stellification of their young Prince" in celebration of his birth. Still, beginning in the late sixteenth century, scientific writers employ forms of the word to refer to "the stelliferous beames of the glistening Sun," and the "vault stelliferous" of the heavens. This last example emphasizes the conceptualization of the heavens as a fixed dome, an architectural form. Indeed, to stellify, to fix with stars, implies the creation of a mosaic-like decoration of the interior of a cathedral. The mythical characters who have been stellified are made first into images then painted onto the very solid but, without such poetic embellishment, essentially empty, sphere of the outer heavens.

The architectural evocations of stellification, a process intimately bound with space as place, suggest a similar sounding, but etymologically unrelated word: stelography. Stelography, literally the practice of placing commemorative inscriptions or designs on steles (upright slabs), tables, or pillars, conceived in relation to the universe as a fixed, essentially architectonic, structure, suggests the textual features of this writing on or with stars. A stelliferous vault, the inside of a dome painted to simulate the universe, serves as a direct metonymy of the universe itself. But such writing, decoration, embellishment, is, further, transformative in nature. The appearance of the glass temple in Book I of *The House of Fame* strengthens this claim. Robert Edwards sees the journey and temple in Chaucer's poem as paramount towards understanding how, for Chaucer, "language creates and preserves texts within an economy regulated by natural law."[19]

To return to the assembly of the classically stellified, the story of Aesculapius resonates with this transformative property. After Aesculapius, a physician, succeeded in restoring the dead to life, and thus overturning the natural order, Jupiter, at Pluto's request, "struck the bold physician with lightning, and killed him, but after his death received him into the number of the gods."[20] Aesculapius transforms the natural order: though he meets his own fate as a result, his stellified form attests to the mutability of nature. After his death, the heavens proclaim his victory. So far in this discussion, we've been following Pluto's peregrinations, paths well engraved by a leaden chariot, alternating between underworld and universe. Another dweller of the underworld, Somnus, the brother of Death (Mors), also figures prominently in our discussion of stellification as writing and movement. But Somnus has no chariot except the stylo employed by the dreamer after dreaming. The stylo (in Greek, *stylos*, a pillar) also implies construction, the creation of architectural forms.

The dream, a vehicle like the chariot, also implies a technology capable of surpassing natural boundaries. Indeed, *techne*, art or artifice, involves the construction of ideas that overcome natural limitations. Further, the dream, with its highly technical vocabulary of interpretive schemata and hierarchies of truth claims, is as technical in its design as Alain's chariot. The dream book, then, serves as a manual for understanding and operating this other vehicle.

ON THE DREAM BOOK

Medieval books of dream interpretation, like medieval bestiaries and lapidaries, represent the encyclopedic qualities of the medieval mentality. As Pierre Brunel notes in *Histoire de la Littérature Française*,

> la mentalité médiévale, profondément symbolique, aimait voir dans chaque animal, dans chaque pierre, le symbole d'une vertu ou d'un

> vice; aussi, dans ces ouvrages, l'édification morale l'emporte-t-elle sur l'information scientifique.
>
> the medieval mentality, profoundly symbolic, loved to see in each animal, in each rock, the symbol of a virtue or of a vice; likewise, in these works, moral edification had the upper hand over scientific information.[21]

Indeed, while dream interpretation resonates as an important component of modern psychoanalysis, "In a sense, Freud, Jung, and others were not so much innovators as restorers, since they were reassigning to dreams and dream-readings the importance that they had held in antiquity, and which they had lost in more recent centuries."[22] For writers of the high and late Middle Ages, dreaming evoked anxiety and fascination; these writers carefully organized the images given to the dreamer in an encyclopedic manner.[23]

There are three different kinds of dream book. While each provides the means for a reader to uncover the meaning of a dream, these three kinds of texts are organized in different ways and require different methods of reading. The "dream alphabet" or "chancebook" "consists of a list of potential dream significations keyed to the letters of the alphabet. The dreamer divines the future by means of a chance process unconnected to the dream's specific content."[24] In this process, the dreamer prays, then opens any book at random. The meaning of the dream corresponds to the signification of the first letter on this page. Thus, the chance book consists of a list of dream images and their potential attendant meaning. But the chance book also provides a method of reading that can be extended to other texts (like the Bible, for example), so that any text, when opened at random, can potentially provide an accurate dream interpretation. However, the actual dream of the dreamer is not very important with this type of reading. Dreamers could also look through the book to find interpretations of specific images, but the reading method of the chancebook emphasizes this random connection between a meaning and the dream. Some sample lines from a chancebook include the following:

> Aerum serenum videre lucrum significat
> (To see a clear sky signifies gain)
> Intestina sua videre secreta manifesto
> (To see one's intestines means secrets revealed).[25]

The second kind of dream book, the "dreamlunar," is like the first in that the actual content of the dream is of little importance. In the dreamlunar, the phase of the moon determines the meanings of dreams. Thus, as Kruger remarks, "on any given night, all dreams predict the same outcome."[26] However, the dreamlunar, linked as it was to the calendar, also mentioned specific days which could greatly influence dreams: manuscripts warn of forty dangerous days "which the masters of the Greeks have tested by experiment," "bromatic days" from the twenty-fourth of November to the eighteenth of December, and "perentalic days" from the first of January to the first of March."[27] Furthermore, they warn of "the days when the leaves fall from the trees,"[28] an action which apparently distorts the clarity of dreams.

The last type of dream book, the "dreambook proper" "simply provides a list of the consequences that will follow from a variety of possible dream contents," so that to dream of losing teeth signifies the death of a kinsman.[29] Each type of dream book is affiliated in some way with orthodox Christianity. The dream alphabet and the dreambook proper are linked, quite directly, to Joseph and Daniel. As Kruger notes,

> the dream alphabet calls itself, in some manuscripts, the *Sompnile Joseph*, claiming one of the two most famous Old Testament dream interpreters as its author. [. . .] In a similar way the dreambook proper claims Daniel as its author, calling itself the *Somniale Danielis*.[30]

Thus, the dream text becomes absorbed by the hermeneutic apparatus of Christian mythography. Regardless of the issue of authorship,

> le Pseudo-Daniel représente, dans la littérature méditerranéenne et occidentale du Moyen Age, un cas singulier par son extraordinaire fortune auprès du public, durant près d'un millénaire. [. . .]Derrière les quelque 200 cas de songes évoqués et à travers une symbolique très primitive, se cache toute la tradition orale du monde païen et de la Bible, autant sans doute qu'aucune source savante.

> the pseudo-Daniel represents, in the Mediterranean and western literature of the Middle Ages, a singular case due to its extraordinary popularity with the public for nearly a thousand years. The entire oral tradition of the pagan world and of the Bible is hidden behind a very primitive symbolism in the nearly two hundred examples of dreams evoked.[31]

This text constitutes, Berriot argues, "un document capital pour qui veut connaître la symbolizue, la mentalité, l'imaginaire même de l'Occident entre le XIIe et le XVIIe siècle" [a key document for whoever wants to know the symbolism, mentality, the imaginary itself of the west between the twelfth and seventeenth centuries.][32]

The structure of these manuscripts also suggests the affinity between the act of reading and the act of dreaming: one manuscript of a dream book, for example, is called the Experiments of Daniel.[33] Perhaps this title hints at the more far-ranging interests of medieval dream interpreters. These texts do not really require a dream to be "read." Instead, they operate as experiments for producing meanings. Indeed, Thorndike mentions a manuscript that is made up of two texts, the *Imago mundi* of Honorius of Autun and the *Re philosophia* of William of Conches, which have been cut up and intermixed with each other.[34] The resulting manuscript evokes a Borgesian text as labyrinth, its paths of possibility expanding into the divine space of the dream. Indeed, Borges' own *Libro de Sueños* suggests this connection. Borges begins his *Libro de Sueños*, a collection of diverse materials on the subject similar to Borges' *El Libro de los Seres Imaginarios*, his paean to medieval bestiaries, with a citation from Joseph Addison, taken from an essay in the *Spectator* (Sep. 1712). Borges notes that Addison "ha observado que el alma humana, cuando sueña, desembarazada del cuerpo, es a la vez el teatro, los actores y el auditorio" [has observed that the human soul, when sleeping, disembarks from the body, and is, at the same time, theater, the actors, and audience.][35] From Addison's observation, Borges arrives at the thesis, "peligrosamente atractiva, de que los sueños constituyen el más antiguo y el no menos complejo de los géneros literarios" [dangerously attractive, that dreams constitute the most ancient and not least complete of the literary genres.][36] Borges' thesis achieves a powerful resonance when considered in relation to these dream books. The intercutting of manuscripts, the variability of possible reading strategies they invite, suggest that such books are not authoritative catalogues, but attempts to recreate the peculiar logic of that very same "más antiguo y el no menos complejo de los géneros literarios." While Kruger divides the various texts based on the reading strategies they require, these reading strategies also hint at an interest in chance and textual order that extends beyond the purely occult interest in dream divination.

THE HIERARCHY OF DREAMS

The cosmological dream narrative is closely aligned with these handbooks on the interpretation of dreams. Macrobius' *Commentary on the Dream of Scipio* is perhaps foremost in bridging the gap between these two genres because it provides both the narrative unfolding of a dream as well as a significant work of dream interpretation. In Chapter III of Macrobius' *Commentary on the Dream of Scipio*, Macrobius provides a catalogue of the various types of dream. As William Harris Stahl notes in his translation of Macrobius, "the elaborate classification and description of dreams [. . .] was one of the most popular sections of the Commentary and caused the author to be regarded as one of the leading authorities on dreams during the Middle Ages."[37] Macrobius' classification of dreams made him one of the preeminent sources on dream interpretation in the Middle Ages.

His classification, however, is not without its own influences. Indeed, as Stahl notes, "the bulk of it bears striking resemblances to the classification given by Artemidorus at the opening of the *Onirocriticon* and at times would serve as a free translation of the Greek work."[38] The *Oneirocriticon* of Artemidorus is, along with Macrobius' commentary, often considered the chief work on the interpretation of dreams that has remained from the Roman Empire.[39] Unlike Macrobius' commentary, Artemidorus' work is more of a handbook for the interpretation of specific dreams. Consisting of five books, the *Oneirocriticon* features lists of things seen in dreams and offers interpretations of these items within the context of the dream. The first three books

> can be regarded as a basically unified and structured treatise on the interpretation of dreams, the last two books certainly have the appearance of a slightly perfunctory study manual. [. . .] The fourth and fifth books [. . .] were intended to be read only by [Artemidorus'] son, also named Artemidorus, who appears to have been a novice in the field of dream interpretation.[40]

The 1603 edition of Artemidorus also features works of dream interpretation by Astrampsychos and Nicephorus.[41] This interspersing of texts in a single manuscript was not unusual, however. The method of arrangement in these books was typified by an alphabetical arrangement of verses describing the meaning of things seen in dreams. While Daniel and Joseph were the primary Biblical figures connected to these books, they were not the only figures with dream books attributed to them.[42] Astrampsychos, for instance, was supposed to have been one of the Persian Magi.[43]

While the *Oneirocriticon* was regarded as a classic on the subject of dream interpretation, Latin authors quoted more frequently Macrobius' discussion of dream hierarchies. Macrobius' hierarchy of dreams is based, most fundamentally, on their possible claims to truth. This distinction of true and false dream was, however, indebted to Greco-Roman mythology. While accurate dreams came from the gate of horn, those without significance came from the gate of ivory. Macrobius clearly points out the validity of this distinction;

he himself cites venerable authority in order to verify the existence of these portals:

> Someone may take the occasion to inquire why false dreams are allotted to the gate of ivory and trustworthy ones to the gate of horn. He should avail himself of the help of Porphyry, who, in his *Commentaries*, makes the following remarks on a passage in Homer presenting the same distinction between gates: 'All truth is concealed. Nevertheless, the soul, when it is partially disengaged from bodily functions during sleep, at times gazes and at times peers intently at the truth, but does not apprehend it; and when it gazes it does not see with clear and direct vision, but rather with a dark obstructing veil interposed.[44]

Furthermore, he points out the rationale for these motifs. The gate of horn presents true dreams because "the nature [of horn] is such that, when thinned, it becomes transparent."[45] The composition of ivory, however, "is so dense that no matter how thin a layer of it may be, it remains opaque."[46] Likewise, Macrobius' presentation of the gates of dreams reveals something else about his schemata. In Greco-Roman mythology, dream images, whether they are true or false, are sent from or filtered through one of these gates. In other words, the dream is still a sort of divine transmission, regardless of its status as true or false. Artemidorus, on the other hand, "cautiously avoids the question of whether dreams are sent by the gods or whether they are motivated by something that is within the dreamer."[47] We can compare this with the dream schemata of a Neoplatonist like Gregory the Great (c. 540-604) who analyzed and categorized dreams in his *Moralia in Job*. Gregory the Great's more elaborate hierarchy divides dreams according to their origins:

> It is important to realize [. . .] that dreams come to the soul in six ways. They are generated 1. either by a full stomach, 2. or by an empty one, 3. or by illusions, 4. or by our thoughts combined with illusions, 5., or by revelations, 6. or by our thoughts combined with revelations.[48]

Their truth values are determined by such origins. Those that originate as a result of physical stimuli are regarded as untrue and ultimately, unreliable.

Macrobius locates five kinds of dream within the dualistic structure of horn and ivory. The *oraculum*, *visio*, and *somnium* are all to be regarded as "true." The *insomnium* and *visum*, on the other hand, are false. As in Gregory the Great's classification, Macrobius aligns the *insomnium* and *visum* with causes within the individual. However, unlike Gregory the Great, who limits these to physical causes, Macrobius indicates that the *insomnium* may be caused "by mental or physical distress, or anxiety about the future: the patient experiences in dreams vexations similar to those that disturb him during the day."[49]

Furthermore, in describing the *insomnium*, Macrobius cites Virgil in order to emphasize the extent to which, even these dreams, untrue and caused by "physical distress," are still emanations from the underworld. Thus, Macrobius notes

> Virgil, too, considers nightmares deceitful: 'False are the dreams (insomnia) sent by departed spirits to their sky.' He used the word 'sky' with reference to our mortal realm because the earth bears the same relation to the regions of the dead as the heavens bear to the earth.[50]

The *visum*, "of no assistance in foretelling the future"[51] occurs in the moments between sleep and wakefulness. As Kruger notes, "the *visum* begins to move beyond the self, suggesting, if only faintly, the transcendence of the purely mundane—the contact with a spiritual (spectral) realm—that will characterize higher, revelatory dreams."[52] Indeed, the *visum* is marked by the appearance of the incubus, "which according to popular belief rushes upon people in sleep and presses them with a weight which they can feel."[53] While Kruger points out that the spectres and incubi of the *visum* "arise [. . .] from 'imagination,' from a misconstruction of reality,"[54] Macrobius does not clearly indicate the origin of the incubi. Aquinas and Albertus Magnus, for instance, would affirm the existence of the incubi as an evil spirit.

Still, this ambiguity highlights the degree to which, in Macrobius' schema, the dream is sent from an external, presumably supernatural, source. It would seem logical to sketch a linear movement away from dreams sent by supernatural sources. But, this is not always so easy to do. Macrobius finds all dreams sent from a supernatural source: the gates of horn and ivory located in the Greco-Roman underworld. From this theorization of dream derivation, Neoplatonists sketch their own schemas of dream types that point out the physical cause of the unreliable dreams. We can see a clear mistrust of the visions created by the habits of the intemperate as well as a validation of the spirit over the body in this sort of arrangement.[55] The continued relevance of Macrobius' division of dreams still influences and dominates ideas of dreams held by medieval philosophers.

The idea that dreams were without spiritual cause was not created by the Neoplatonists, however. Aristotle's own statements on the validity of dreams had traditionally been used to deny the supernatural origin of dreams. Calcidius, in his *Commentary on Plato's Timaeus*, positions his own consideration of the diversity of dreams against "Aristotle [. . .] who [. . .] would dismiss all divination and deny that future things may be known."[56] He did not deny Aristotle. Instead, he based his syncretic theory of dreams on an attempt to discover truth in the concept of dreams espoused by both Aristotle and sources like Heraclitus and the Stoics who acknowledged the mystical derivation of dreams. However, Calcidius concludes that "neither [approach] is the exclusive explanation of the question. He finds the solution to the problem of reconciling the conflicts of the two opposed theories in Plato, for whom the 'ratio somniorum' is 'multiformis.'"[57] As a source influencing theories of dreaming, Aristotle was believed by medieval commentators to view all dreams as mundane in origin. While his *Parva naturalia* does include statements that attest to the supernatural origin of some dreams, "Aristotle elaborates a theory in which dreams are essentially internal phenomena, caused by the interaction of psychology (sense perception, imagination) and physiology (the movement and purification of blood attendant upon the processes of eating and digestion)."[58]

The higher dreams that Macrobius mentions come from the gate of horn.[59] Authoritative knowledge, though reflected from a divine source, is reflected through "a dark obstructing veil." Macrobius does not spend as much space describing the qualities of the three varieties of true dream as he does the untrue or unreliable varieties. Macrobius describes both the *insomnium* and *visum* from verses three to eight of Book Three of the *Commentary*, despite the fact that he is most concerned with explaining Scipio's "true" dream. His description of the *oraculum*, *visio*, and *somnium*, on the other hand, occupy a much shorter space (this from verses nine to eleven in Book Three). In the *oraculum*, the dreamer receives instruction from some authoritative figure. The *visio*, which "comes true"[60] seems to indicate a premonition which is not narrated or revealed. Instead, the dream image is presented in some manner akin to how it will be experienced.

The *somnium*, or enigmatic dream, "conceals with strange shapes and veils with ambiguity the true meaning of the information being offered, and requires an interpretation for its understanding."[61] The definition is not particularly clear. Instead, it emphasizes ambiguity; also, unlike the definitions Macrobius provides for other types of dream, there is no clear example of the possible content of the *somnium*. Indeed, the *oraculum*, for instance, is defined by its content: we can identify it by the presence or absence of an oracular figure. Likewise, we are unsure of the relation between the *somnium* and waking reality. This is not the case with either the *oraculum* or *visio*, however. Instead, in the *oraculum*, the prophetic actor in the dream clearly connects the dream world and the waking world. In the *visio*, this connection is not made immediately apparent. Instead, the dreamer will be aware that the dream was a prophetic vision once the event it prophecies comes to pass.

In the *somnium*, however, these connections are obscured. Despite this, Macrobius tells us that "we need not explain further the nature of this dream since everyone knows from experience what it is."[62] Further, he offers five additional varieties of *somnium*: personal, alien, social, public, and universal.[63] These categories move progressively outward in a manner akin to Stephen Daedalus' address in *A Portrait of the Artist as a Young Man*[64]:

> It is called personal when one dreams that he himself is doing or experiencing something; alien, when he dreams this about someone else; social, when his dream involves others and himself; public, when he dreams that some misfortune or benefit has befallen the city, forum, theater, public walls, or other public enterprise; universal, when he dreams that some change has taken place in the sun, moon, planets, sky, or regions of the earth.[65]

These levels within the *somnium* complicate classifications of dreams between the main categories of dreams. They seem to encompass any possible type of dream. But Macrobius does not seem to be concerned with erecting impermeable boundaries between the different categories of dreams.[66] The dream of Scipio itself "embraces the three reliable types mentioned above, and also has to do with all five varieties of the enigmatic dream."[67] Thus, while the original title of the work (*Commentarii in Somnium Scipionis*) implies that Scipio's dream is only a *somnium*, it is also an *oraculum* and a *visio*. This is interesting to note because we customarily view such a hierarchy as built of mutually exclusive categories. The medieval philosophers building from Macrobius certainly contribute to this method of reading the dream levels.

Even Kruger's discussion of Macrobius emphasizes "the realm of dreams as split between the true and the false, the predictive and the deceptive."[68] Kruger goes on to suggest that Macrobius has described the "realm [of dreams] as a hierarchy, a graded system that proceeds steadily from one extreme term to another."[69] Kruger provides a chart of Macrobius' categories and arranges the five categories from higher to lower possibilities of truth, moving from the *oraculum* and *visio* as completely true to the *somnium*, which is truth in fiction, to the *visum* and *insomnium*, both of which are false.[70]

As I said, this arrangement is clearly not unusual. Medieval readers of Macrobius were inclined to view dreams and other natural mysteries in this manner. The reason I am pointing this out, though, is because, given the actual text of Macrobius' commentary, we can see that Macrobius is not at all concerned with such rigid levels of "truth." If he were, then Scipio's dream could not have both the absolute truth claim of the *oraculum* and the "intermediate" truth claim of the *somnium* which, according to John of Salisbury, "stretch[es] before the body of truth a curtain [. . .] of allegory."[71] Macrobius also uses this trope of truth covered by a veil of allegory in his discussion of the two types of *narratio fabulosa*. Here, Macrobius distinguishes the form usable by philosophers by its "decent and dignified conception of holy truths, with respectable events and characters [. . .] presented beneath a modest veil of allegory."[72]

Macrobius' text sets out a structure which seems logical; it implies a hierarchy based on the relative truth value of possible types of dreams. But Macrobius' own interpretation of the dream of Scipio does not seem that rigidly connected to the seemingly mutually exclusive categories of dreams. Macrobius' inclusion of Scipio's dream into the five varieties of the *somnium* could, of course, reveal a highly complex narratological schema. In this sense, Macrobius recognizes that each part of the dream must be categorized separately. To regard Macrobius' classification this way is perhaps to think of the dream as a heteroglossic narrative made of distinct narrative strands.

In Book Three, he does divide the narrative into five separate parts, and then demonstrates the connection between each part and one of the five varieties of *somnium*. However, such a possibility would not have resonated with medieval commentators precisely because of the fact that, for these commentators, the truth value of a dream was of the utmost importance. As Jon Whitman remarks in *Interpretation and Allegory: Antiquity to the Modern Period*, "The efforts of Christians to interpret Plato, Macrobius, Martianus Capella, and others according to Christian doctrine already acquire extensive expression (including detailed textual 'glosses') in the early Middle Ages, and by the twelfth century these efforts take boldly sustained forms in the elaborate commentaries and cosmological reflections of Christian philosophers and theologians."[73] In other words, the reception and use of the auctores by medieval commentators was closely tied to an interweaving of pagan philosophy and Christian faith. This particular combination, intended to give intellectual value to Christian philosophy and spiritual credence to pagan philosophy, animated much of the writing by Christian commentators. The form of the encyclopedic compendium of knowledge that was so favored by medieval commentators was not intended solely to enumerate the phenomena of the natural world. Instead, such works also valuated natural phenomena within a spiritual framework. Thus, perhaps insecure about the intellectual foundations of Christian philosophical writing (when compared to the tradition of pre-Christian sources), Christian commentators appropriated these sources by magnifying or accentuating their cohesiveness.

DREAMS AND DREAMS—AS—NARRATIVES

In Macrobius, the underlying ethical rationale for the dream differs quite markedly from that possible by medieval Christian commentators. Besides the origin of dreams which we have looked at, we must also consider the purpose of dreams. The purpose of dreams as established by Macrobius must also force us to rethink the hierarchy of true and false espoused by, for example, Augustine's similar discussion of dream interpretation. At the beginning of Chapter Four of the *Commentary on the Dream of Scipio*, Macrobius remarks that "the purpose of the dream is to teach us that the souls of those who serve the state well are returned to the heavens after death and there enjoy everlasting blessedness."[74] We know that he is referring to Scipio's dream, and that, as in the previous chapter, he is engaged in an act of dream interpretation. However, the boundaries between dream interpretation and hermeneutical analysis become blurred. A dream is an actual phenomenon. It is random, unplanned, and unscripted by the dreamer. The categories that Macrobius presents in Book Three are clearly intended to describe just this phenomenon of the dream. Texts, regardless of their claims to divine inspiration, do not whisper through the gates of horn and ivory.

However, Macrobius' comment that the dream has a purpose, and that purpose is essentially to provide a didactic message regarding the virtues of civil service, brings us back to the *Commentary on the Dream of Scipio* as a text. Here, the essentially pre-Christian focus of Macrobius' *Commentary* is at its most poignant. He is, after all, discussing a text which forms the concluding section of Cicero's *De re publica*, itself indebted to Plato's *Republic*.[75] In these texts, dreams appear as attempts to maximize the rhetorical effect of carefully reasoned arguments. Indeed, as Ludwig Schrader notes in his discussion of Joachim Du Bellays' "Songe" (1558), the dream vision did not record an actual unconscious state and, instead, relied on the imitation of other dream visions: "Du Bellay verwendet selbst, ich zitierte eine solche Stelle, den Ausdruck 'imitation.'" [Du Bellay himself used, I give here one such passage, the expression 'imitation.'][76] Artemidorus likewise comments on this distinction between the dream and its representation, noting that "Some mention should also be made of the fact that certain things that are seen in dreams are there only for the sake of embellishment and that one must leave these details out of his interpretation."[77]

One of the most obvious examples of the apparent overlapping of the category of dream as actual experience versus the dream as a rhetorical position intended to have a didactic purpose comes from the content of dreams provided by classical and medieval writers. Lucretius' *De Rerum Natura* provides one example of a dream catalogue that makes a direct correspondence between the social position of the dreamer and the dream as a single, plotless extended image. His conception of the dream space parallels his mechanical model of the universe. For Ralph Crum, author of *Scientific Thought in Poetry*, "Lucretius is, indeed, unique [. . .] because he bends all his efforts towards making a mechanical conception of the universe acceptable to his readers, on the ground that such an explanation frees man from fear and superstition and makes him master of the forces of the universe, for through the power of his mind he can control his will."[78] The similarity between the mechanical universe and mechanical unconscious lends itself to a conceptualization of people and things as agents, their zombie-like personalities determined exclusively by their actions.

Indeed, in discussing the allegorical consciousnesses animating Lucretius' *De Rerum Natura*, Newman notes that

> the mind returns in dreams to its habitual interests, its special delights, its customary employments: the lawyer pleads cases, the sailor battles the winds, generals direct battles, libertines greedily watch swaying dancing girls, the child searches for a chamber pot (and finds when he wakes that he has wet his bed), an adolescent embraces a lover in a spasm of lust—and Lucretius himself philosophizes in his sleep.[79]

The dreams in this example are all single actions. Likewise, these hypothetical dreams tend to turn the dreamers themselves into dream-images. By designating a dreamer as a "sailor," for instance, Lucretius posits that dreamer as a signifier capable of only limited signification. Thus, the sailor dreams of battling the winds. The sailor can only dream of battling the winds, every night, for as long as he is a sailor. The limits of the dreamers are likewise implied in these hypothetical dreams. The examples of the libertine, the incontinent child, and the masturbatory teenager reveal dreamers who are trapped by the weaknesses of their bodies. Their dreams are decidedly physical in their manifestations. The same could also be said of the sailor and the general, whose professions are reliant on physical mastery of the elements and hostile forces.

This battle against the body is not specific to these examples, however. The rhetorical strength of these examples comes from the allusions they make to Lucretius' own summation of his dreaming experiences. Unlike those ruled by their bodies, Lucretius dreams of philosophy. We recall, of course, that the medieval and classical hierarchies of dreams, based on truth values, are customarily arranged according to the extent to which these dreams are negatively impacted by the body. Lucretius implies his own mental mastery through this apparent catalogue of a psychological phenomenon.[80] Thus, the dreams mentioned operate more as dreams as examples in an argument aimed at the validation of the writer's own argumentative skills. There is a clear connection between these dreamers who appear, at first glance, to be entirely unrelated. The libertine, adolescent, and child do battle against themselves, their physical impulses, in the dream. The sailor and general battle the outside world. Furthermore, the word battle is used to evoke both the sailor and the general. By extension, Lucretius successfully battles the physical world both externally and internally, gaining for himself a dream space of the highest order.

The purpose of the dream, then, fits with the text in which it appears. The fabulous narrative, in this case the dream itself, is a suitable form for the discussion of government. The reason I am emphasizing the textual nature of the dream is because this is of key importance for the hierarchy of true and false which medieval commentators extrapolated from Macrobius' discussion of the dream of Scipio. Macrobius designates the design and purpose of the dream with the term *skopos*. Again, he focuses on the inherently textual quality of the dream despite the fact that his hierarchy of dreams is concerned with dreams as psychological phenomena and not as texts.

Let us return to Kruger's chart judging the relative truth values of the five kinds of dream. We must be aware of the correspondence between texts and dreams. While Macrobius was often used as an authority on dreams as non-textual psychological phenomena, his text does not make such clear divisions between the two categories. Indeed, as we have seen, he affirms that Scipio's dream is a text. It is a text presented in the form of a dream. Further, it is a *narratio fabulosa*. While the dream of Scipio may lay claim to truth because, as dream, it originates from the gate of horn, its claim to truth as a text is more suspect. For, as Macrobius reminds us, the *narratio fabulosa* is clearly inappropriate as a form for any type of discourse on "the Highest and Supreme of all gods, called by the Greeks the Good (*tagathon*) and the First Cause (*proton aition*)."[81] Thus, a paradox seems to exist. While it may be possible to have a vision with an unimpeachable claim to truth because it originates from a divine source, one could say that the actual "truth" of this passage is called into question when it is represented in the form of narrative.

In other words, the form and content of representation have a deep, intrinsic connection. While it is relatively easy to conflate dreams and narratives, to posit dreams as narratives, to apply the "rules" for dream interpretation to textual interpretation and/or vice versa, the two are clearly distinct. There are, in fact, separate criteria for evaluating the truth value of each. Thus, Macrobius reminds us that "when [. . .] philosophers speak about [. . .] the Supreme God and Mind, they shun the use of fabulous narratives."[82] Here, Macrobius merely recapitulates a point that stemmed from Plato's own writings and was likewise reaffirmed by Neoplatonists like Abelard.[83]

The injunction against the use of fabulous narratives for the explication of the *tagathon*, *proton aition*, or the *nous* was also linked to a conception of the communicability of absolute knowledge that stretches into the contemporary attitude toward the communication of verifiable concepts found in the sciences. Peter Machamer's "The Nature of Metaphor and Scientific Description," for example, offers a discussion of the linguistic limits and assumptions that accompany the use of metaphor in the creation of scientific concepts. He notes the interplay between metaphor and description, pointing out that the boundaries between these apparently distinct rhetorical modes are not as seamless as often imagined. Early in his discussion, Machamer notes that

> in older Aristotelian terms, a definition, one type of description, displayed the essence of something by placing the defined thing into a genus (and by providing the *differentiae* of that thing compared to other things in the genus). We need not be concerned with essences, but with sorting things by types or kinds; sorting into categories and into subcategories. So categories not only select things by kinds and by common groupings, but often bring all of a given kind under a wider category. It is in this way that categories are normally hierarchical.[84] However, he moves from this fairly standard summarization of Aristotelian hierarchies to assert that "hierarchies are not all the same."[85]

This is an important point for my own discussion because, as I have suggested, the kinds of hierarchies that are so evident in the medieval discussion and application of the levels of dreams formulated by Macrobius and Calcidius also reveal the limits inherent in such systems of organization. Indeed, Machamer goes on to explain that

> sometimes things are ordered hierarchically by part-whole relations, other times in terms of composition, and yet other times by types, properties or instances. The whole forms a categorical network, and what needs to be noted here, is that these networks may exhibit many different types of relations. The categorical structure gives the ontic types, or displays the kinds of things and properties, that are basic for a domain.[86]

Machemer moves from hierarchies to an analysis of their role in the formation of models used to describe scientific principles; however, his discussion applies to both the scientific and the scientistic. A prime example in his own essay, Ann Noble's "Wine Aroma Wheel" becomes the centerpiece of the problem of standardizing the gradations used to measure the experience of a sensory phenomenon. Indeed, while Machamer calls the study of wine, or oenology, "my favorite science"[87] one could argue that the subjectivity of the evaluative component makes it, while a worthwhile endeavor of course, more scientistic than scientific.

Machamer concludes that "what one wants to call metaphors in science are always just descriptions, and they function just as descriptions always do."[88] We will return to this point, as it illuminates the extent to which the description-laden form of the cosmological dream allegory contributes to the nascent scientific discourse that emerges in Kepler's use of this form.

Absolute knowledge, because it was linked to the Divine, was unrepresentable.[89] However, the quest to represent absolute knowledge was of key importance: philosophy and theology are, above all, attempts to understand the Divine. Still, the Absolute was never entirely removed from the world of lived experience. While the Ten Commandments, for example, warn against the evils of forging images of God, and many prophets turned their heads from the overpowering and blinding face of a God revealing His power, Adam and Eve were fashioned in the image of God. Furthermore, the New Testament, fulfilling the covenant between God and humanity through the Incarnation, attests to the expression of the Divine in material forms.

Medieval audiences were well aware of the complexities of this interrelationship. Visual iconography reinforces the literary poetics I trace here, a process apparent in works such as Jan Van Eyck's *The Adoration of the Mystic Lamb*. The central figure of the altarpiece, either Jesus or some combination of the aspects of the trinity, is depicted in a regal, but human form. His human form is the same as that of Adam, Eve, Mary, and John the Baptist, all of whom flank Him. In the bottom panels, however, Christ and the Holy Spirit are depicted in their zoomorphic incarnations: the mystic lamb and the dove of the spirit, respectively. These animal forms of the various aspects of the Christian God are not intended to be read as the protagonists of an animal fable.[90] The gold lettering on the altar suggests a method of "reading" the lamb: *Ecce Agnus Dei qui tollit peccata mundi.* But an interpretation of this phrase involves more than a recognition that Christ the lamb takes away the sin of the world. Instead,

> Hier hebben we te maken met een transpositie van het ritueel van de zondebok, die op de grote Verzoendag symbolisch beladen werd met de zonden van Israël, en weggestuurd in de woestijn (Lev 16, 20-22). In het Nieuwe Testament werd het thema van de zondebok overgedragen op de verzoening door het Lam Gods. Jezus heeft immers door het offer van Zijn leven alle zonden van de wereld weggenomen.
>
> Here we have to make a transposition from the ritual [of sacrifice] of the Old Testament where the Great Day of Atonement was symbolically tainted by the sins of Israel, who had been exiled in the desert (Lev 16, 20-22). In the New Testament, this theme from the Old Testament was overtaken by the reconciliation of the mystic lamb. Jesus has eternally overcome all of the sins of the world with the offering of his life.[91]

The lamb and dove, while not representations of sacrificial animals, were also not meant to be read literally as images of God. In this way, the animal images are more easily thought of as idea-images, particularly when compared to realistic depictions of a human Jesus. Late-medieval and Renaissance artists aimed, above all, to achieve depictions of Christ that, in their physical realism, would convey the spiritual depths of Christ's Passion.

Van Eyck's *The Adoration of the Mystic Lamb* is an example of the complexities inherent in navigating form and representation. If we read this altarpiece as a narrative, we are struck by this opposition between the *narratio fabulosa* and the intrinsically philosophical quest to depict truth.[92] This relates to my argument because the dream narrative as genre is itself subject to this same apparently irresolvable conflict.

Indeed, Van Eyck's use of perspective and placement of figures in the bottom panels of the painting create a dream-like experience. By this, I mean that they create a narrative that is not viewed from an exterior position. Instead, they create a narrative that envelops the viewer. The optical illusion I am referring to involves the empty space between the two main groups of figures adoring the mystic lamb. This space, immediately before the fountain flowing with waters of the Holy Spirit, creates the illusion of inclusion: it is as if the viewer is also part of the painting.

Spectrophotographic analysis of the painting reveals that the fountain, which blocks the path between the viewer and the mystic lamb, was not part of the original underpainting. In contrast, the upper panels of the painting depict figures from an exterior standpoint. While the realism of these panels, particularly of the choir, is frequently cited as testament to the artistic achievement of Jan Van Eyck, these figures are still viewed from an exterior point that is clearly outside of the painting. Of course, a viewer can not actually be "in" a painting. But the use of perspective in the bottom panels blurs the boundaries between inside and outside to an extent not evident in the upper panels. These bottom panels may not actually depict a dream, but it is not too difficult to identify these images as part of a mystical vision.

In depicting a mystical vision, Van Eyck is less confined by the limits of realistic representation. By this, I do not mean that the bottom panels do not demonstrate the same attention to realistic detail which characterizes all of his work. Instead, I mean that the aspects of the trinity are depicted in forms which link them more closely to their meanings than to their "actual" form. Jesus was not, in any way, conceived of as some kind of magic sheep capable of discharging blood at will. This image, however, conveys the philosophical subtleties of the crucifixion and the sacrament of communion. The image depicted as dream aspires to the same level of truth accorded the divinely-revealed dream, or mystic vision.

The dream and the dream as depicted in narrative are, however, two intrinsically different things. When depicted in narrative form, the dream immediately becomes a *narratio fabulosa*. Indeed, the dream also becomes a self-consciously applied rhetorical strategy. We can not forget that the dream of Scipio, as it appears in Cicero's *De re publica*, operates as a deliberate rhetorical move to emphasize, through a formal juxtaposition to the rest of the work the merits of involved civic duty.

As *narratio fabulosa*, however, the dream, for Neoplatonists, enters into a murky layer of dubious authenticity. However, the dream as a rhetorical device also provides the ability to demonstrate concepts that could not otherwise be covered. Thus, in discussing Plato's conclusion to the *Republic*, Macrobius asks "But how could Plato show that [one's enjoyments] continued after death except by demonstrating the immortality of souls?"[93] For Macrobius, the dream narrative with its depiction of the shape of the afterlife follows quite logically from the already reasonable and logical anatomy of a society: "After he had created a belief in the immortality of souls, he drew the obvious conclusion that the souls, upon being released from their bodies, had definite places allotted them according to their deserts."[94]

The *narratio fabulosa* was not, then, condemned outright as unsuitable for any attempt at philosophy. The various categories of the fable that Macrobius mentions are useful for comparing the dream narrative and the actual dream. The *narratio fabulosa* is, in Macrobius' schemata, at an interesting juncture of truthfulness and falsehood that is not that different from the claims of truth of the dream itself. Indeed, like the categories used to classify dreams, the *narratio fabulosa* is subject to the same hierarchy of forms based on truth values. The *narratio fabulosa* is, above all, only one kind of fable. As a form, it is then, *a priori*, subject to skepticism. As Macrobius reminds us, "the very word acknowledges [the] falsity" of fables.[95] However, while false, the fable can also inspire the reader to good works. The two main categories of the fable are divided by their purpose. One type is intended merely to gratify the ear. Macrobius disparages this category, relegating such fables to "children's nurseries."[96] This kind of fable, exemplified by Apuleius's *Golden Ass*, provides purely aesthetic pleasure.

The second kind of fable, divisible into two subcategories, conveys virtue. The first subcategory can be characterized by fictitious setting and plot. Macrobius provides Aesop's fables as key examples of this category. The second, the *narratio fabulosa*, "rests on a solid foundation of truth, which is treated in a fictitious style."[97]

There are, however, two types of *narratio fabulosa*. The first type is distinguishable in that its plot "involves matters that are base and unworthy of divinities and are monstrosities of some sort."[98] This type of fable, though based on "a solid foundation of truth," is not suitable for philosophy primarily because it includes texts which feature the "gods caught in adultery, Saturn cutting off the privy parts of his father Caelus and himself thrown into chains by his son and successor."[99]

The second type of *narratio fabulosa* differs considerably from the first and lends itself to the "decent and dignified conception of holy truths."[100] This category certainly includes the cosmological dream allegory, treating as it does the true structure of the cosmos and its connection to philosophical truth within the fictional veil of dream. As a literary category, it most closely parallels the dream category of the *somnium*. In this sense, both categories are somewhere between true and false in that they shroud the truth within a fictional veil. Thus, this version of the *narratio fabulosa* "is the only type of fiction approved by the philosopher who is prudent in handling sacred matters."[101] However, this type, as we have seen, cannot aspire to treat "the Highest and Supreme of all gods."[102]

The form cannot support the most important sacred matters because the *tagathon*, *proton aition*, and *nous* lie beyond the realm of language itself.

To a certain extent, no literary form is capable of representing that which lies beyond language. The specific prohibition against the *narratio fabulosa* must, then, lie in some other quality of the form. For one, perhaps it is the only form that aspires to represent the unrepresentable. This very quality, however, links the *narratio fabulosa* and the dream book to the nascent field of astronomy. Another important reason for this critique of the form lies in the similarities between the boundaries of language and the boundaries of the universe itself. Thus, the two kinds of celestial vehicles meet in the form of their representation. The process of stellification, simultaneously mythographic and scientific, impels the identification of the boundaries of language with the boundaries of the universe itself. As I mentioned earlier, the model of the universe presented by Macrobius and appropriated by Neoplatonist medieval commentators shows the material world divided from the world of the spirit through a series of interlocking spheres. This same model of material reality reflects the attitude toward linguistic representability reflected in Macrobius' *Commentary on the Dream of Scipio* and was widespread through the Middle Ages. The very categories which Macrobius declares unrepresentable are etched on the surfaces of the celestial spheres.

Chapter Five

John of Salisbury's Critique of the Dream Book

MATHEMATICA AND *MAGICA*

Attributed to Daniel, works designed to aid in dream interpretation enjoyed considerable popularity among Medieval and Renaissance audiences.[1] The dream itself operated as a space where the supernatural intersected with the personal. The numerous manuscripts which provided catalogues of motifs and possible interpretations claimed their authority from the Bible, giving the genre a legitimacy denied to many classical texts that were only recuperable through dubious testaments to the Christian leanings of authors.[2]

While Christian philosophers were struggling with the creation of a philosophy that was Christian, or a Christianity that allowed for philosophy, in the *Polycraticus* John of Salisbury, a Christian philosopher, denies the possibility of a science built from Christianity. Through his critique of the dream book, John of Salisbury both suggests the nascent scientificity of discourse in the dream books and asserts the diabolical derivation of this discourse. In this sense, the dream books represent a step toward the development of a modern scientific discourse. The logical processes required by the dream books as texts suggest an attempt to systematize the dream as a natural phenomenon. At the same time, the interpretive process of the dream books suggests correspondence between the natural world and spiritual thought. This viewpoint animates much of early astronomy as evidenced, for example, by the extent to which Kepler's theories stem from such seemingly bizarre themes as the tonal harmony of the spheres. This impulse in science does not, however, die out with modernity. Gerald Holton's discussion of Henri Poincaré's resistance to the theory of relativity in *The Thematic Origins of Scientific Thought,* for instance, stresses the continued relevance of spiritual questions on attempts to understand the natural.

For medieval audiences, the very process of dream interpretation was problematic. Dreams changed meaning depending on outside conditions. The natural world impacted the supernatural space of the dream. The elements and the stars impacted a dream. The dream interpreter was reliant on the authority of works such as the *Somniale Danielis* as well as the interpreter's own knowledge of the movement of the stars. This intermingling of the interpretation of the dream and the interpretation of the cosmos yields the development of modern astronomy. By this, I do not mean only that the quest for more accurate dream interpretations impelled the early astronomer. Instead, the very foundation of the *astronomicus* had associations with the *diabolicus* that are perhaps best understood as expressions of the poetic imagination. Astronomical knowledge thus coextends with the reshaping of the theory of allegory as both are shaped by the medieval dream narrative.

John of Salisbury's *Polycraticus,* as our main example in this chapter, includes a trenchant critique of both the *auctoritas* of the *Somniale Danielis* and the validity of the very practice of dream interpretation. For Salisbury, writing in the twelfth century, an interest in dream interpretation was not limited to mediums or soothsayers on the fringes of society. Instead, the process was closely linked to the interests of the most learned. John of Salisbury was certainly warning against the most egregious misuse of dream books; as a teacher interested in solidifying the intellectual claims of a uniquely Christian philosophy, he was warning against practices which would delegitimize or appear in direct contradiction to Biblical authority.[3] In doing so, John of Salisbury presents the etymological similarities of the terms *magica, mathematica,* and *maleficium,* and concludes that the interpretive processes necessitated by the Dream Books as genre imply an understanding of God's order which is unachievable except through divine inspiration.

More specifically, John of Salisbury's critique of the *Somniale Danielis* raises significant questions about the connections between *magica* and *mathematica* as legitimate means of discovering knowledge. These connections are frequently invoked by historians of science interested in contextualizing the development of astronomy, as a modern science based on logic and reason, from astrology, an occult practice with no claims to logic or reason. Here, I am interested in these connections from a narratological standpoint. What do the narrative structures of dream books and oneiric narratives reveal about reading practices and the means available to medieval writers interested in the correspondences between the book of the world and the visions of the dreamer?

In the section of the *Polycraticus* dedicated to dream interpretation, John begins with a description of the harms associated with physical spectacle. The examples he gives of deleterious physical spectacle involve music and theatrical shows. Like music and theatrical shows, harmful *artificia* like the magic arts and *mathematica* "arose from a fatal familiarity of men and demons."[4] Physical spectacle, for a Christian scholar, would serve as a kind of short-hand for the diabolical, or at least blasphemous, because of its suggestion of idolatry, of the veneration of physical desires above spiritual piety. The term *artificia* suggests that which is made by and venerated by man. Salisbury's choice of music and theatrical shows as points of comparison for his critique of the magic arts and *mathematica* hints at the dark worldliness of desires to understand the physical world. On the most basic level, Salisbury's choice of music, for example, suggests a worldly sensuousness anathema to a Christian philosopher, and this despite the extent to which "Patristic allegoresis makes the Muses harmless through euhemeristic explanations and reinterprets them as concepts in musical theory."[5] But theories of music, emphasizing the natural order of tonality, were often invoked as ways of understanding the presence of God, construed as a force of divine order, for the idea of the music of the spheres, for instance, guided, or at least motivated, the works of medieval thinkers heavily influenced by Christian theology.

This indictment of the magic arts and *mathematica* demonstrates the close links between moral philosophy, nascent scientific investigation, and the arts, especially those of not just language, but also spectacle or theater. By linking *mathematica* and the magic arts, Salisbury suggests the sinful nature of nascent scientific disciplines. Theatre, like music, suggests spectacle in ways that are not as clearly aligned with sin as possible, however. Perhaps unknowingly, John of Salisbury uses an example of spectacle that is, as Fletcher points out, linked to the cosmological concerns of the medieval mathematician. His use of theatre as a trope intended to highlight the dangers of *mathematica* reveals instead the extent to which the two are interconnected at a fundamental linguistic level.

Indeed, the spectacle and the spectral spectacle of overhanging constellations require the same attention from their observers. The spectacle, as that which is observed from an outside standpoint, reflects on social phenomena exterior to the performance as a narrative embedded within a culture outside of that narrative. Like the medieval astronomer, seeking miasma and cholera in the movements of characters like Orion and Ursus, the audience of the play watches the spectacle in order to decode events outside of the play. As long as Salisbury's critique is only limited to the actions of astrology, a non-science, or to the magic arts of divination, it retains its value from the standpoint of promoting a Christian philosophy which is both moral and astute. However, when we consider the links between astrology, dream interpretation, and the eventual development of an astronomy recognizable as a science in the modern sense, then we can view John of Salisbury's connection of terms such as *magica, mathematica,* and *maleficium* as expressive of a divide between philosophy and science which is, perhaps, bridged by the use of allegory.

John of Salisbury's indictment of *artificia* also testifies to his own advancement of rhetoric, a discipline not so distant from medieval astronomy. His defense of rhetoric likewise takes the tenor of this contrast between the natural and unnatural. In reaction to the works of a Cornificius, "who regards rhetoric as superfluous and undertakes to philosophize without it," John asserts that

> Rhetoric is the beautiful and fruitful union between reason and expression. Through harmony, it holds human communities together. He who would put asunder what God has brought together for the good of men deserves the name of public enemy ('hostis publicus'). For to take Mercury from the arms of Philology, to eliminate rhetorical theory from the study of philosophy, is to destroy all higher education ('omnia liberalia studia').[6]

Concepts central to astronomy also informed rhetoric. The term *kosmos,* for instance, was understood as a totality that reflected the attributes of the mind of God. The same term, however, also refers to the blatant ornamentation of the allegorical image. As Fletcher notes, "the oldest term for ornamental diction, '*kosmos,*' appears in Aristotle's list of the eight types of words that constituted, as he saw it, poetical language."[7] The idea *kosmos* is buried within and itself disguised beyond recognition in Latinate derivatives such as *ornatus* and *decoratio.* Indeed, "the etymological connections of decorum and decoration, polite, police, and expolitio, cosmic and cosmetic, costume and custom, with all their minor variants (e.g., 'ornamental gardening,' 'proper dress') all demonstrate the same fundamental duality."[8] The duality of the term suggests the difficulties of translating or representing a physical reality in a mimetic form. The physical reality is, as its denotations of macrocosmos and microcosmos attest, concerned with an item of the utmost importance: namely, life, the universe, and everything. But the importance of the representations of this totality in a visual or verbal symbol does not carry over. Instead, the representations carry a negative connotation which is at the center of critiques of allegory as a narrative practice. As *ornatus* and *decoratio,* the allegorical image does not merely represent the *kosmos;* instead, it overrepresents what is already a comprehensive totality. This process renders the allegorical image superfluous and thus highly artificial.

The same could be said for the *narrative* creation of the allegorical image. The critiques of the allegorical image stem from its ostentatious presence, and the extent to which it suggest the fullness of the meaning it represents through an accretion of detail. This aspect of allegory is, perhaps, linked to the process by which allegorical images produce meaning. But this process is also linked to meaning in an emblematic or occult sense, and indicates a process of interpretation where the significance of the image is greater than the image itself. The term *kosmos*[9] then constitutes the most fundamental type of allegorical image. Signifying as it does both "(1) a universe, and (2) a symbol that implies a rank in a hierarchy"[10] it refers to both a macrocosmic body that engulfs all individual images, a world devourer that is, itself, a total image made of all other images. This sense of allegorical creation parallels Gordon Teskey's thesis regarding the violence inherent in allegorizing. It also denotes a specific image or *specula* embedded within and giving meaning to this greater macrocosmic whole.

For John of Salisbury's medieval audience, the various terms linked to *mathematica* held a range of significations perhaps lost on the contemporary reader. The word *mathesis,* as John points out, denotes learning in general. However, when it has a long penultima, it signifies a process based on divination. As John warns us, magic itself is difficult to define cohesively. Magic can be "many and diverse;"[11] thus, the boundary between magic and dream interpretation remains obscured by the tacit Biblical approval of dream interpretation.[12] While passages of the Bible do warn against divination, the prominent example of David's successful interpretation of the dream of Nebuchadnezzar affirms, from a doctrinal standpoint, the extent to which divine aid can result in the interpretation of a dream.

NARRATIVE TIME AND STRUCTURE OF DANIEL'S DREAM

In the example of David and Nebuchadnezzar, David succeeds in both receiving and interpreting another's dream. For Nebuchadnezzar, these are not distinct practices; instead, telling and interpreting "are one in his mind."[13] He does not merely interpret the king's dream. Instead, "the mystery was revealed to Daniel in a vision."[14] With this first dream, the story emphasizes the extent to which Daniel, unlike the "magicians, enchanters, sorcerers, and astrologers"[15] of Nebuchadnezzar, can intuit the content of the dream and interpret the meaning of the dream. The king enacts a severe penalty for failure to fulfill both operations for this dream:

> The king replied to the astrologers, 'This is what I have firmly decided: If you do not tell me what my dream was and interpret it, I will have you

> cut into pieces and your houses turned into piles of rubble. But if you tell me the dream and explain it, you will receive from me gifts and rewards and great honour. So tell me the dream and interpret it for me.'[16]

This process embeds the king's dream, as meaning and content, in a further vision, or *somnium.*

Daniel receives the king's dream, entirely disembodied from its dreamer, in a separate vision; the king's dream, then, constitutes a vision within a vision. If Nebuchadnezzar's dream were a narrative, Daniel's vision would encapsulate that narrative both within the narrative frame of Daniel's own dreaming and also within the interpretive apparatus (perhaps most easily imaginable in the textual form of footnotes) that characterizes the divine origin of Daniel's own vision. At this point, Nebuchadnezzar's dream is as disembodied as possible. The text removes the dream from the physical contingencies of the dreamer, for example.[17]

However, at this point of ultimate disembodiedness, of distance from the physical world, we are confronted by a physical object. The dream relies heavily on *ekphrasis,* a term defined by Murray Krieger as "the picture-making capacity of words in poems."[18] In the dream, Nebuchadnezzar stands before "a large statue—an enormous, dazzling statue, awesome in appearance. The head of the statue was made of pure gold, its chest and arms of silver, its belly and thighs of bronze, its legs of iron, its feet partly of iron and partly of baked clay."[19] There is action in the dream, but the agent of the action remains disembodied as well. Thus, "while you were watching, a rock was cut out, but not by human hands. It struck the statue on its feet of iron and clay and smashed them."[20] Or, if anything, it is the rock that is the agent of the dream. The dream ends as "the rock that struck the statue bec[o]me[s] a huge mountain [that] fill[s] the whole earth."[21] The dream could be succinctly summarized as the tale of a rock crushing the feet of a statue, triggering a chain reaction that destroys the entire statue.

This summary focuses more on the action of the dream than on the objects in the dream. Daniel's interpretation, though, relies heavily on decoding the physical forms of the dream: these include the statue and the rock. While I commented earlier on the extent to which Daniel's feat of dream interpretation results in a dream that is incredibly abstract and removed from physical reality, the dream itself can be narrowed to objects that are not just material objects, but imposing or sublime material objects: a giant statue and a rock the size of a mountain. Furthermore, the duration of Daniel's oneiromancy is linked to his ability to describe the objects of the dream. His interpretation of the dream begins in verse 36 and ends with verse 45. Of these ten verses, eight are concerned with describing and interpreting the meaning of the statue. Above all, Daniel's revelation is tied to the materials used to construct the statue.

Any consideration of duration in this particular narrative is itself subject to the problems of reception and embedding that characterize the dream. The distinction between *erzählte Zeit* (story time) and *Erzählzeit* (narrative time) is further problematized by the reception of the dream by both Daniel and Nebuchadnezzar, for example. These concepts refer to temporal disjunctions between the time of the story (*erzählte Zeit*) and the time of the narrative (*Erzählzeit*). Genette notes that while the opposition between *erzählte Zeit* and *Erzählzeit* is typical of cinematic narrative, it is also a constitutive component "of oral narrative, at all its levels of aesthetic elaboration, including the fully 'literary' level of epic recitation or dramatic narration."[22] In film or oral narrative, the notion of *Erzählzeit* is fairly self-explanatory: a film is a certain length; a story told takes time in the telling. However, written narratives seem to avoid these problems until we consider that the written literary narrative, "like the oral or cinematic narrative, [. . .] can only be 'consumed,' and therefore actualized, in a *time* that is obviously reading time."[23] The two main categories of anachronies help us to conceptualize temporal distinctions. Prolepsis designates "any narrative maneuver that consists of narrating or evoking in advance an event that will take place later" while analepsis refers to "any evocation after the fact of an event that took place earlier than the point in the story where we are at any given moment."[24]

Given the double-dreamed nature of this vision, we would be at a loss to clarify the duration of the dream-itself, particularly given the difficulty of determining the actual dreamer-as-author of the Biblical text in question. Daniel is most certainly the speaker[25] of the dream, and thus could be described as its author. His language creates a powerful image which he then unveils, revealing its further significance. However, as the dream's narrator, he also refers to the presence of the king in the dream. Daniel's narration begins with reference to the king, thus creating an additional narrative level. The king dreams; later, Daniel intuits its content; later still, Daniel narrates the dream; within this narration, the point of view is filtered through the perceptions of the king, and not of Daniel. The king "looked" at the statue; he is the observer. Daniel, on the other hand, merely narrates from an external, but unidentified, position.

The distinction between levels of narration suggests the complexities of interpretation necessitated by dream interpretation as an extratextual practice. I mean here that we must consider the Dream Book and the Book of Daniel which impelled the creation of the Dream Book as genre as a continuum of text and analysis which never loses sight of phenomenological reality as text. The process of dream interpretation is, instead, closely linked to events outside of the dream and beyond the text. The duration of the dream as told by Daniel differs markedly from the dream as experienced by Nebuchadnezzar. As Genette indicates, we can consider events in relation to the dual axes of analepsis and prolepsis provided by the concept of anachrony. In this text, we are confronted by a disjunction between anachrony, achrony, and interpretation. The order of the story, or the *erzählte Zeit,* follows a very simple pattern, which we can divide into seven sections: Section A ("You looked, O king, and there before you stood a large statue"), Section B (Description of the statue), Section C ("While you were watching, a rock was cut out"), Section D ("It struck the statue on its feet of iron and clay and smashed them"), Section E ("Then the iron, the clay, the bronze, the silver and the gold were broken to pieces at the same time and became like chaff on a threshing-floor in the summer"), Section F ("The wind swept them away without leaving a trace"), and Section G ("But the rock that struck the statue became a huge mountain and filled the whole earth").

The order of the narration follows the same pattern. There is a direct correspondence between the order of events and their narration. However, as I noted earlier, there is a great disjunction between the amount of narrative time spent on recounting the significance of these various events. Likewise, we must consider the correspondence between Daniel's narration of the events and images of the dream and of their meanings. As I demonstrated earlier, direct interpretation characterizes a large majority of Daniel's presentation of the dream. His interpretation follows the order of the dream in a linear sequence. First the dream is presented. Then, the dream is decoded, image by image, in the order in which each was presented. However, this order seems in direct contrast to the varied and variegate varieties of order and duration provided by Genette on the same subject. Genette's choice of the *Iliad* and of Proust as two ends of an implied history of narrative suggests that we can anticipate a great number of anachronies in nearly any narrative.

The concepts of narrative time proposed by Genette illuminate the general structure of the dream narratives in Daniel. However, the narrative methods employed are also dependent on the genre which they represent. Erich Auerbach's distinction of the Homeric and Biblical narrative, for instance, stresses that these two approaches stem from entirely different purposes and worldviews. Genette, however, provides examples from Homer in his anatomy of Proustian narrative. The distinctions between these two genres (Homeric and Biblical narrative) speak to their relationship to allegory. Genette presents the narrative features from a standpoint that is unconcerned with the overall generic structure of the text. For Auerbach, there is no way to read either text without being aware of the far-reaching psychological implications of each approach to narrative. Thus, these distinctions help to clarify the relationship between narrative features and genre in Daniel. The interpretations of dreams as reported in Daniel (and which influenced subsequent books of dream interpretation) rely on *ekphrasis* and result in an extended consideration of the allegorical image. The dream narrative and its interpretation, revealed to Daniel by the divine, represent a *kosmos* in terms of both meanings mentioned by Angus Fletcher. The time of the telling suggests, more than anything, a scanning of the object which is featured in the dream. These features of the recount of Daniel's dream interpretation contribute to the form of the medieval dream book.

Auerbach distinguishes the two "poles" of western literary influence through the externalization of phenomena in the Homeric epic and the internalization of mental states in the Biblical narrative. In discussing digression in Homer, Auerbach points out that

> The excursus upon the origin of Odysseus' scar is not basically different from the many passages in which a newly introduced character, or even a newly appearing object or implement, though it be in the thick of a battle, is described as to its nature and origin; [. . .] indeed, even the Homeric epithets seem to me in the final analysis to be traceable to the same need for an externalization of phenomena in terms perceptible to the senses.[26]

The genre of the epic is here connected to the numerous passages which depart from the sequence of the narrative in order to describe a person, place, or thing. This is a result of the external quality of the epic. The original cause for this externalization of sensory phenomena "must have lain in the basic impulse of the Homeric style: to represent phenomena in a fully externalized form, visible and palpable in all their parts, and completely fixed in their spatial and temporal relations."[27] The epic is, as described here by Auerbach, an essentially pictorial mode. The externalization of emotions and mental states becomes translated into the externalization of objects, and even of the transformation of events in the narrative into objects that are described by the narrator.

The narrative as description characteristic of the epic relies on the relationships between textual foreground and background. The movement from foreground to background constructs or maintains suspense in the epic. The creation and maintenance of suspense remains the key problem concerning

Auerbach in his discussion of Odysseus' scar. While the Homeric epic features an apparently more sophisticated wealth of narrative techniques than the Biblical narrative, it does so at the expense of the clearly defined suspense found in Biblical passages such as the story of Abraham and Isaac. Thus, "an episode that will increase suspense by retarding the action must be so constructed that it will not fill the present entirely, will not put the crisis [. . .] entirely out of the reader's mind." Instead, "the crisis and the suspense must continue, must remain vibrant in the background."[28] This is part of a discussion in which Auerbach refers to the interruption of one event by the recollection of another. Here, he refers to anachrony in a manner similar to that provided by Genette. However, Auerbach introduces these anachronic aspects of the narrative because of their connection to a thing. That thing is introduced in the title of Auerbach's chapter: "Odysseus' Scar." The digression accompanies this thing "not [. . .] to keep the reader in suspense, but rather to relax the tension."[29] Digression as a technique does not, for Auerbach, play a key role in Biblical narratives. These are interested primarily with the simple narration of events. Tension is achieved through the linear recount of simple, but significant series of events.

The main distinctions Auerbach draws between the Homeric and Biblical narratives are, indeed, concerned with the narrative structuring of time, the inclusion and maintenance of suspense, and the presence of the pictorial in the narrative. My evocation of Auerbach follows my application of Genette's concepts of anachrony to Daniel's dream interpretation because I want to demonstrate the connection of time and digression in the dream narrative as represented in Daniel. The space and time of the dream in Daniel are still narrated in an orderly linear fashion: first the dream is recounted, then the interpretation is provided. However, as far as narrative time is concerned, the dreams themselves deal with objects. There is no real linear narrative to the dream. The dream (and here I am referring to the prophetic dream) represents a world in its entirety through the creation and contemplation of a single allegorical image. This process is key to John of Salisbury's critique of the works of dream interpretation that stem from Daniel.

We must not lose sight of the purpose of the dream books. Auerbach distinguishes Homeric epic and Biblical stories on the basis of psychological factors. The great strength of his argument stems from the extent to which the philosophy of a culture is embedded in the aesthetic form of its narrative output. Thus, the teleological impulse of Jewish (and, ultimately, Christian) religion and worldview becomes mimetically modeled in the narrative of Abraham taking Isaac up to the mountain. The element of suspense remains, although it is not linked to any mystery unrevealed to the reader or the characters of the narrative. Both Isaac and the reader are aware of where the story is "going" when Isaac asks "The fire and wood are here [. . .] but where is the lamb for the burnt offering?"[30] Indeed, Genesis 22 begins with God's command that Abraham "take your son, your only son, Isaac, whom you love, and go to the region of Moriah. Sacrifice him there as a burnt offering on one of the mountains I will tell you about."[31] The maintenance of suspense reveals a cultural belief in signs and the meanings of things orchestrated by a divine cause.

Homeric epic, however, specifically avoids allegory. Indeed, there seems to be no concern with an "outside" of the narrative. By "outside" of the narrative, we can refer to an external cause of the events narrated or the external narrative apparatus (footnotes, hermeneutical analyses) necessary to interpret a work. In the discussion of Abraham and Isaac, Auerbach is concerned with the extent to which suspense leads outside of the narrative. It does this by leading us to a recognition of God's ability to effect change in the material world. The ram which Abraham finds on the mountain and offers as a gift to God has already been given by God to Abraham. On another level, this narrative serves to clarify the bond between God and the Jewish people that is likewise embedded in this story of sacrifice. The suspense of the intended audience of this text (Jewish and then Christian) is not merely a concern with Isaac's fate. Instead, Isaac's fate and the possibility of his survival becomes an indication of the fate or possibility of survival of the audience as well.

This movement outside the narrative is also of key importance in the presentation of the dreams of Nebuchadnezzar. The dreams themselves are presented as allegories. Daniel reveals the images of Nebuchadnezzar's dreams, then decodes these images. Like the narrative of Abraham and Isaac, this passage reveals the influence of God in the world. Daniel emphasizes the extent to which his powers of oneiromancy stem from his connection with God. God reveals and interprets Nebuchadnezzar's dreams to Daniel. Again, the narrative moves towards an exterior vantage point. That vantage point, the ability to Asee" a dream and to decode it, comes only from a position close to the divine.

DREAM INTERPRETATION AS DISCOURSE

But it is not just the narrative itself which causes problems for John of Salisbury. Instead, it is the dream as narrated in Daniel which influences, in John of Salisbury's view, the slippery slope toward daemonic divination. In the *Polycraticus,* John goes on to speak of the various relations between the actor of a dream and the audience of interpretive reception. The potential complexities of this are clear given the context of dream interpretation provided by Daniel. John of Salisbury does not refer to these complexities in order to credit Daniel's seemingly divine powers of interpretation. Instead, John of Salisbury does this to discredit any interpretive function provided by oneiromancy:

> Somnium [. . .] gerit imagines, in quibus coniectorum praecipue disciplina versatur, et nunc suum cuiusque est, nunc alienum, modo commune, interdum publice aut generale est. Ut enim ait Nestor, de statu publico regis credatur somnio.

> Now the dream concerns the dreamer himself, now someone else, now common interests, sometimes the public or general welfare. And, after Nestor, trust is put in the king's dream concerning public matters.[32]

The acts of dream interpretation in the Bible are not linked to an individual's own perception of self; the Biblical dreamer does not dream of his relationship with God in and of itself. Instead, such dreams are closely linked to the common interests of the Jewish people, for example. Or, as in the books of the prophets, the visions revealed by God to the prophet are intended for the edification of all people.

Finally, as both the first and second dream in the book of Daniel attest, the king's dream is an accurate precognition of the status of public affairs. Nebuchadnezzar's dream of a tree results in events that impact both the king and his kingdom. For Nebuchadnezzar, the negative consequences of the fulfillment of the dream are obvious: "You will be driven away from people and will live with the wild animals; you will eat grass like cattle and be drenched with the dew of heaven."[33] The status of the kingdom is impacted by the king's eventual recognition of the God of Israel. This chapter in Daniel features passages that are directly narrated by Nebuchadnezzar. The character of the relationship between the king and God is concealed within this dream which operates as a "decree the Most High has issued against my lord the king."[34] While the first-person narration of these dream passages emphasizes the direct relationship between the dreamer and the Divine, the political implications of the dreams indicate that the interpretation of a dream is intended to benefit the entire kingdom.

For John of Salisbury, however, this extension of the dream past the confines of the dreamer hints at a process which could be divine, but is much more likely a result of diabolical agents. After clarifying that oneiromancers often consider the individual dream as a boundary-less transmission that can impact any number of people, he moves to the variables which impact interpretation. In summarizing dream interpretation as practiced by his peers, "he explains that the season of year when one dreams, the place where one dreams, and the personal characteristics of the dreamer must all be taken into account; that sometimes interpretations should be by contraries, and again from like to like."[35] These criteria of interpretation suggest a hermeneutical framework which lays great claims to empiricism (despite the ephemeral and non-material character of dreams). Time, place, and the physical and psychological status of the dreaming subject are all taken into account by the conscientious oneiromancer. The image John provides of medieval dream-interpretation seems to presage contemporary practices of psychoanalysis, which lays great claims to the scientific verifiability of the influence of these factors on an individual's mental state. We could even graph these factors of time and place of the dream onto diachronic and synchronic axes respectively. However, John of Salisbury is not persuaded of the cohesiveness of the body of knowledge created by dream interpreters. Instead, he notes that:

> Sed dum has coniectorum traditiones ex(s) equimur, vereor ne merito non tam coniectoriam ex(s) equi, quae aut nulla aut inania ars est, quam dormitare videamur.

> But while we pursue these traditions of the interpreters, I fear lest we deservedly seem not so much to trace the art of interpretation, which is either no art at all or an idle one, as to dream ourselves.[36]

He suggests that there is no cohesion in the theory and texts underlying the work of the dream interpreter. However, John of Salisbury's estimation of dream interpretation and its relation to the diabolical demonstrates the complexity of questions concerning reason and faith and their influence on empirical knowledge of the world.

The dream book, a reference book outlining the meanings of common dream motifs, was designed to allow the reader to uncover divine significance in his own life. Furthermore, biblical authority affirmed the legitimacy of the interpretations provided by the books. The dream book, then, served to bridge the linear historical model of correspondence and premonition characteristic of Christianity's salvation timeline and the history of the life of the individual. While actual miracles, or direct contacts between God and man, may have been reserved exclusively for Biblical persons and the saints, the dream book provided a reasoned catalog of the visual devices God could use to speak to man in the narrative space of the dream. The texts seem to link an interest in the empirical with the spiritual despite John of Salisbury's warning that:

> Verum quisquis credulitatem suam significationibus alligat somniorum, planum est quia tam a sinceritate fidei quam a tramite rationis exorbitat.
>
> Whoever fastens his credulity to the significations of dreams evidently wanders as far from sincere faith as from the path of reason.[37]

The dream book as text was, for John of Salisbury, clearly outside of the boundaries of both faith and reason. But he also seems to deny validity to the signification of dreams in any way. The phase "signifactionibus alligat somniorum" does not necessarily mean only "those significations provided by the dream books." Instead, John of Salisbury casts dreams as devoid of significatory value. Or, to a greater extent, he views the dream as a diabolical device that promises meaning and syntactic cohesion, but instead leads the dreamer into disarray and disorder. His reference to "wandering" and "the path of reason" turns the dream into a will o' the wisp, a spectral presence leading unwitting travelers to their deaths.

The existence of these dream books suggests as well a different connection between biblical events and people's own lives than that suggested by Auerbach. The form of representation of Biblical narratives, examined by Auerbach through the question of realism, demonstrates the Biblical narrator's own investment in the truth of his work. The Biblical narrator represented that which a culture believed to be true:

> The Biblical narrator was obliged to write exactly what his belief in the truth of the tradition (or, from the rationalistic standpoint, his interest in the truth of it) demanded of him—in either case, his freedom in creative or representative imagination was severely limited; his activity was perforce reduced to composing an effective version of the pious tradition. What he produced, then, was not primarily oriented toward 'realism' (if he succeeded in being realistic, it was merely a means, not an end); it was oriented toward truth. Woe to the man who did not believe it![38]

In this passage, Auerbach identifies a narrator who, given the context of this discussion, is only the narrator who depicts Abraham and Isaac. But the overall claim that Auerbach makes regarding the realism of Genesis seems to extend to all Biblical narrators. Auerbach implies that the attitude toward historical truth which he postulates for the narrator of Abraham and Isaac's journey also serves to describe the attitude toward realism of the many other Old Testament narrators. For the story of Abraham and Isaac, this makes sense; the question of 'realism' versus truth resonates because of the obvious intrusion of the supernatural into the realm of human affairs. The narrative is, for the most part, concerned with whether or not the supernatural will act through the physical world. Truth, arrived at through faith, is at odds with the real because of the unverifiable status of the supernatural. However, in the dream sections in Daniel, we see people responding to dreams and visions: the appearance of dreams, while mysterious, is not at odds with the rules governing the natural world.

The truth claims made by Biblical texts suggest a process of historical truth. As religious texts, they deal with the supernatural; but the supernatural operates as an agent engaged in nation-building. The purpose of the Bible as a vessel of historical truth results in an urgency which drives the texts:

> One can perfectly well entertain historical doubts on the subject of the Trojan War or of Odysseus' wanderings, and still, when reading Homer, feel precisely the effects he sought to produce; but without believing in Abraham's sacrifice, it is impossible to put the narrative of it to the use for which it was written. Indeed, we must go even further. The Bible's claim to truth is not only far more urgent than Homer's, it is tyrannical—it excludes all other claims. The world of the Scripture stories is not satisfied with claiming to be a historically true reality—it insists that it is the only real world, is destined for autocracy.[39]

As Auerbach stresses, the world of the Scripture insists that it is the only real world. Thus, all phenomena described within the texts also describe, by extension, these phenomena beyond the Biblical texts. My claim here is certainly not revolutionary: note the tendency of some contemporary Christians to still view Genesis as the last word on the development of natural ecosystems and species differentiation. An investigation of the ramifications of that dispute would necessitate an entirely different study.

But I think this distinction between the historical and physical world illuminates my consideration of medieval dream narratives and cosmological narratives. First, the boundaries between these two categories (historical and physical) were not as clearly defined for a medieval audience. Second, the connection between the historical and personal are key to Christian philosophy. Third, the dream space represents a space which connects, on the one hand, historical and physical reality and historical and personal lived-experience on the other.

The element of interpretation, whether applied to events or physical realities, serves to further unite the historical and the physical. In fact, the connections between the historical and physical impel interpretation:

> Doctrine and promise are incarnate in [Biblical texts] and inseparable from them; for that very reason they are fraught with 'background' and mysterious, containing a second, concealed meaning. In the story of Isaac, it is not only God's intervention at the beginning and the end, but even the factual and psychological elements which come between, that are mysterious, merely touched upon, fraught with background; and therefore they require subtle investigation and interpretation.[40]

Here, Auerbach summarizes the fact that Biblical texts, unlike Homeric epic, demand interpretation. The texts are, as well, expressions of a world outside of the text. This world, like the text, is mysterious, touched by the Divine, and requires interpretation. To this extent, "doctrine and the search for enlightenment are inextricably connected with the physical side of the narrative—the latter being more than simple 'reality'; indeed they are in constant danger of losing their own reality, as very soon happened when interpretation reached such proportions that the real vanished."[41] Consider the distinction Auerbach makes here. Doctrine and the search for enlightenment are connected to the claims to physical reality made by the narrative. Because of this relationship between doctrine/ search for enlightenment and physical reality, the former contribute to the "super-natural" qualities of the natural. Doctrine and the search for enlightenment help to shape the physical side of the narrative, leading to the synthesis of narrative depiction of an event and the layers of textual and "actual" significance which constitute the *figura.*

But Auerbach also stresses in this passage that interpretation can overtake a text in such a way that the process of interpretation becomes disentangled from its subject. Interpretation overtakes the text, perhaps in a manner similar to Kepler's *Somnium* where the footnotes outweigh the actual narrative. Thus, the Biblical narrative "seeks to overcome our reality: we are to fit our own life into its world, feel ourselves to be elements in its structure of universal history."[42] This accounts for the powerful influence of Biblical narratives on the shapes and patterns of order imposed on western history. However, this process which begins with interpretation in the service of universal history "becomes increasingly difficult the further our historical environment is removed from that of the Biblical books."[43]

These connections between Biblical history and the natural world become more difficult to maintain as different methods of explaining the natural gain significance:

> as late as the European Middle Ages it was possible to represent Biblical events as ordinary phenomena of contemporary life, the methods of interpretation themselves forming the basis for such a treatment. But when, through too great a change in environment and through the awakening of a critical consciousness, this becomes impossible, the Biblical claim to absolute authority is jeopardized; the method of interpretation is scorned and rejected, the Biblical stories become ancient legends, and the doctrine they had contained, now dissevered from them, becomes a disembodied image.[44]

These claims to authority manifest themselves in the various attitudes toward the dream books expressed in the late Middle Ages. John of Salisbury's critique of the dream book as genre demonstrates that there was a wide attachment to the interpretive possibilities it promised. Furthermore, the intersection of the dreamer and the Biblical clarify the extent to which it was possible to "represent Biblical events as ordinary phenomena of contemporary life." But there is also this matter of an "awakening of critical consciousness" which, in this analysis, threatens the Biblical claim to absolute authority.

DREAMS AND DEMONS

After John of Salisbury attacks the *Somniale Danielis* as a text, he moves to a discussion of dreams in general as the work of demons. The initial alliance between man and demon which he had originally postulated as the diabolical combination of forces which had led to the very production of the *Somniale Danielis* becomes, then, something inherent in the dream itself. Furthermore, John notes the subtle but significant distinctions between the work of divine inspiration and occult art. John uses the book of Daniel to point out that only Daniel, and none of the King's other astrologers, was able to interpret the dream: "But notice that the privilege which man could not confer was given to Daniel alone, to bring to light the riddles of dreams and to scatter the obscurities of figures."[45]

His use of terms such as "riddles" suggests that dreams are solvable. If they are solvable, then the dream is made up of symbols and images which can be diagrammed within a given interpretive framework. The critique of the other astrologers stems from God's assistance in this decoding process.

For John of Salisbury, dreams might be decodable, or may suggest a system of tropes and figures which assigns meanings to events and images taken from the material world. But, the interpretive system of dream images is just as complex as the interpretive system necessary in order to understand the book of the world. This process, likewise, is organized according to a hierarchical model, so that accuracy of interpretation is linked to one's closeness to God. In every way, "medieval Christianity posits a rationally structured cosmic hierarchy whose parts are meaningfully disposed along the ascending way to God."[46] In other words, the links between things and meanings is infinite and, hence, only knowable by the infinite.

I think it is useful to distinguish "riddles" from "figures" as the terms are used by John. The decoding of riddles, or of specific groupings of things, suggests a synchronic catalogue of these things and their meanings. An object and its meaning are always connected in the same way. The dream of a Biblical personage and the dream of a European in the Middle Ages may be decoded with recourse to the same pairings of object and meaning. However, the term "figures" suggests a diachronic dimension to this interpretive process. As figural interpretation, in the Biblical sense, is concerned with correspondences between different periods of history, a decoding of figures by the dream interpreter suggests the prophetic dimension of dreams. Furthermore, this prophetic dimension implies the relevance of the life of the individual within the panorama of Christian history. By this, I mean that, because the Dream Books provide catalogues of interpretations supposedly endorsed by Daniel, the prophetic dreams of individuals are also connected, at the level of the unconscious, to Biblical authority as absolute authority. At this point, we see an evocation of Biblical authority that exceeds Auerbach's formulation. The astrologer, working with methods which, though primitive, anticipate the scientific, builds these methods to correspond with the claims of Biblical authority. John of Salisbury, as a rhetorician, theologian, and philosopher, questions the degree to which Biblical authority actually extends into the lives of his contemporaries. He questions the extent to which astrologers write their own lives into the context of Christian history through dreams:

> Are the interpreters of dreamers thus wont to examine thoughts and remove obscurities, to explain what is involved and illuminate the darkness of figures? [. . .] He whom the spirit of truth does not illume vainly puts his confidence in the art of dreams.[47]

John of Salisbury critiques contemporary dream interpreters because of the vanity and uncertainty inherent in proclaiming that one is illuminated by the spirit of truth. The strength of his argument stems from its reliance on Christian humility: if one says he is illumined by the spirit of truth, then he most certainly is not. However, the interpreters of dreams are doing something different. They are not proclaiming themselves the recipients of divine inspiration: their reliance on catalogues of dream symbols does not suggest oracular activity. Instead, these vast catalogues, covering a wide range of oneiric possibilities, operate as a comprehensive system of knowledge which affirms the presence of Biblical authority in the daily lives of people who are not prophets or saints. While the audience of these dream books may not see burning bushes or be able to endure the heat of a furnace, they can affirm the presence of the supernatural in the dream world which, while not material reality, is still, of course, able to be experienced. For John of Salisbury, on the other hand, dreams themselves are the work of demons:

> Quis huius facti explicet rationem nisi quod boni spiritus vel maligni exigentibus hominum meritis eos erudiunt vel illudent? [. . .] Quod si materiam vitiis afferat, libidinem forte accendens aut avaritiam aut dominandi ingerens appetitum aut quidquid huiusmodi est ad subversionem animae, procul dubio aut caro aut spiritus malignus immittit.[48]

The strength of their incursion into waking life and the forceful images which animate memorable dreams cannot, for John, suggest a purely divine source. Instead, because dreams are spiritual products, they can equally be the work of malignant or benevolent spirits. Furthermore, in the material world, the influence of malevolent spirits outpaces that of benevolent spirits. John's critique separates the Biblical period of history from the historical trends of John and his contemporaries. He repeatedly stresses this disjunction between the world as it was experienced in Biblical time and the very different spiritual composition of contemporary life.

John of Salisbury's indictment of mathematics builds on his conception of the diabolical foundations of the post-Biblical world. After discussing the extent to which dreams are the work of demons, John of Salisbury proceeds to align the demonic origin of dreams with the demonic origin of mathematics:

> Possit utinam tam facile mathematicorum error a praestantioribus animis amoveri quam leviter in conspectu verae fidei et sanae conscientiae istarum illusionum demonia conquiescunt. Verumtamen eo periculosius

> errant quo in soliditate naturae et vigore rationis suum fundare videntur errorem.
>
> Would that the error of the mathematic could be as readily removed from enlightened minds as the works of the demons fade before true faith and a sane consciousness of their illusions. But in it men go astray with the greater peril in that they seem to base their error upon nature's form foundation and reason's strength.[49]

To a modern reader, John of Salisbury's suppositions seem ludicrous. Mathematics, for the modern reader, is the ultimate language of objectivity. To align it with the demonic seems to allow no room for any type of language that is not already corrupted by the demonic. Because numbers are more exact than words, John of Salisbury's own writing could be influenced by the demonic. The snares of rhetoric are more dangerous than the errors of miscalculation.

Still, John attributes to the demons infesting mathematics the same kinds of words used to deny the validity of dreams. The equation of "mathematicorum error" and "illusionum demonia" does not mean that medieval mathematicians and astrologers succumbed to demonic forces merely by calculating incorrectly. Instead, "mathematicorum error" implies errors at the basis of the calculations made by mathematicians. In this sense, it seems as though, to use Gerald Holton's term, John of Salisbury faults the thematic dimension of medieval mathematicians. This poses problems for an understanding of medieval science, however, because, as John has demonstrated, these medieval mathematicians are using Biblical authority to supply the thematic. The dominant thematic of the dream books is, indeed, the logical explanation of the anatomy of dreams found in Daniel.

Thus, although *magica* and *mathematica* are not so much etymologically related terms, as I mentioned earlier, they are, for John of Salisbury, essentially the same thing. Both aim for a control of the natural world that is justified on a supernatural level. Again, the mathematician goes astray and wanders off the path provided by Biblical authority, if we consider the dream book as part of a linear move toward a mathematics that is removed from and not justified by the spiritual. For example, we could read the phrase "in soliditate naturae" as the beginning of a veneration of the laws of nature over the laws of God. Furthermore, if we view "vigore rationis" as referring to a body of knowledge not connected to the Biblical, then perhaps we can view medieval dream interpretation as a move away from the claims of Biblical authority. However, the natural foundations explored in the dream books stem from the supernatural: the catalogue of symbols and motifs are analyzed given their appearance in dreams which are already viewed as a narrative space completely governed by and subject to the workings of supernatural powers. Thus, the natural world, as experienced in dreams and catalogued in dream books, emanates from the divine.

John of Salisbury concludes his critique of the dream books by noting that nature, reason, and experience all fall to error in that man cannot claim knowledge that belongs to God alone. However, the boundaries between the knowledge made by man (mathematics) and that given by God (philosophy or theology) are not as clear as they are often construed. Indeed, Kepler's mystical imagination forces him to confront similar questions in order to reconcile Christianity and Copernicanism.

Chapter Six

The Journey, the Book, and the Dream: An Overview of the *Somnium*

THE JOURNEY

In the previous chapter, I discussed the extent to which twelfth century allegorists fused classical and Christian components in the production of texts at once philosophical and theological. Indeed, for these Neoplatonists, as Steven Kruger indicates in *Dreaming in the Middle Ages,* "dream, soul, and universe are all, in some sense, coextensive."[1] The correspondences between dream, soul, and universe allowed for an accurate because divinely-inspired representation of the cosmos within the narrative space of the dream. However, the model of the universe depicted by such writers is, as is quite obvious to modern readers, scientifically invalid. The universe is not composed of a series of interlocking spheres. The dimensions of the human body do not reflect the shape of the cosmos. Still, the philosophical impulses underlying the creation of such cosmic models are not absent in the formation of those which are more recognizably correct or verifiable.

In the following chapters, I will be analyzing Johannes Kepler's *Somnium,* a work heavily indebted to the Neoplatonist cosmological allegory as a literary form.[2] This relatively minor work of Kepler's has recently become quite popular among literary scholars and philosophers of the history of science.[3] Why has this work garnered so much attention lately? Kepler is, of course, a major figure in the history of science. His newfound popularity among literary scholars does not stem from a recent discovery of Kepler. Along with Galileo and Copernicus, Kepler is one of the triumvirate of the new astronomy. Further, the *Somnium* itself is not a newly discovered work.[4] Marjorie Nicolson, most significantly, established the *Somnium* as a key text for early modern English literature by linking Kepler to Donne and Milton and also through her close attention to the textual complexities of Kepler's writing.

We could attribute the increased significance of this work to the growing influence of literary studies of scientific texts. Such significance portends for the rhetoric of science for English studies, as Marjorie Nicolson attested decades ago. This field of study, interdisciplinary by nature, seeks to address the areas of conjunction and disjunction between texts designated as literary, scientific, or both.

Still, early modern scientific texts exert a certain appeal because of the entirely different conception of disciplinary boundaries. Early science is a discourse indebted to all of the components of the Trivium and the Quadrivium. Kepler's work features a mixture of the medieval arcane and the complexities of modern scientific discourse. It re-imagines the cosmological allegories of Alain de Lille and Bernardus Silvestris while anticipating modern astronomy, embodied in Kepler's own *Nova Astronomia.* This is why it stands at the center of the present study.

Philosopher of science Gerald Holton makes the following remarks about Kepler's own fusion of sources and intellectual concerns:

> The important publications of Johannes Kepler [. . .] ha[ve] been strangely neglected and misunderstood. [. . .] Part of the reason lies in the apparent confusion of incongruous elements—physics and metaphysics, astronomy and astrology, geometry and theology—which characterizes Kepler's work. Even in comparison with Galileo and Newton, Kepler's writings are strikingly different in the quality of preoccupation. He is more evidently rooted in a time when animism, alchemy, astrology, numerology, and witchcraft presented problems to be seriously argued. His mode of presentation is equally uninviting to modern readers, so often does he seem to wander from the path leading to the important questions of physical science.[5]

Kepler appropriates these incongruous elements with a peculiar fluency. His preoccupation with the non-scientific could be said to outweigh, or at least heavily influence, the parts of his work that continue to remain relevant. The Daemon imprisoned at the center of the *Somnium* blasphemes the celestial order of medieval allegory through geometric proofs that resounded with greater force through the seventeenth century.

Indeed, in *Science and Imagination,* Nicolson goes so far as to compare this Daemon with Milton's Lucifer in *Paradise Lost.* This is not the only text that shares the philosophical and theological concerns echoed in the *Somnium,* however. Francis Godwin's *Man in the Moone* (1638) and John Wilkin's *Discovery of the World in the Moone* (1638) both "consider[ed] seriously the possibility of lunar voyages and an inhabited moon, using principles of the new science to buttress their arguments."[6] Further, the scientific writings of Thomas Harriot and William Gilbert also explore the new astronomy. John Donne's *Ignatius His Conclave* (1611) and Ben Jonson's court masque *News from a World Discovered in the Moon* (1611), on the other hand, parodied scientists' "Lunatique" obsession with the moon. Indeed, in *Ignatius His Conclave,* Donne attacks Kepler personally, arguing that "ever since Tycho Brahe's death [Kepler] hath received it into his care that no new thing should be done in heaven without his knowledge."[7]

Despite Kepler's obviously mystical concerns, Kepler has often been presented as a scientist valorizing empirical proof and embracing a proto-scientific method. Such analyses present Kepler as the prototype of the modern scientist. However, as Mary Baine Campbell argues, this line of reasoning ignores Kepler's use of fiction.[8] We can not, then, distinguish Kepler's scientific achievements from the rhetorical form of their textual presentation. Thus, while Alexander Koyré's influential *The Astronomical Revolution* writes Kepler as a Priest of an incipient modernity without whom "the progress of astronomy would have been delayed for a century,"[9] we can not assume that Kepler's achievements came about despite the mystical.

Instead, Kepler remains "an unanalyzable compound of rationality and irrationality" as long as we conceive of the rational and irrational as mutually exclusive categories, or we assume the mystical imagination to be a liability. We must not, then, imagine Kepler as dualistic, a Janus at the gate of two irreconcilable and feuding fiefdoms. Instead, like Argus, the hundred-eyed guardian of Io, identified through the allegorical interpretation of mythology as the moon,[10] Kepler surveys his pale charge through a panoply of constantly telescoping apertures. Further, his *Somnium* is exactly the kind of text that privileges seemingly irreconcilable extremes: it is a composite of rationality and irrationality.

My analysis will show that this composite is interesting to literary scholars because of what it shows us about the creation of distinct discursive modes. In other words, the critique modern historians level against Kepler arises in part because they position him between two discursive poles that were not conceived of as irreconcilable opposites for the late medieval and early modern periods. However, if all of Kepler's contemporaries were equally impacted by the intellectual contingencies of their given era, then why is Kepler an outcast, grudgingly recognized for his achievements but glazed over because of works like the *Mysterium Cosmographicum?* Why is Kepler such a space oddity when compared to Galileo, who is widely historicized as the prototypical man of science, an empiricist who confronts theology and tradition in an ethical quest for an absolute truth unbounded by doctrine?[11] Not all historians subscribe to this view, of course. As Paul Oskar Kristeller stresses in *Renaissance Thought and its Sources,* "Even if we want to say that Kepler discovered the laws in spite of, and not on account of, his Platonist cosmology, as historians we cannot be concerned only with those parts of his work and thought that have been accepted as true by later scientists."[12] Still, for most historians, Kepler, like the moon, merely reflects the light of Galileo, the blazing center of the universe created by the scientific revolution.

It is not possible to attribute Kepler's development of his astronomical theories to this one work, particularly when considering Kepler's vast output. However, Kepler's obsessive devotion to this text suggests that it merits a degree of importance in inverse proportion to its relatively short length. We are, of course, inclined to attribute Kepler's formulation of the theory of elliptical orbits to solid science, to mathematics, graphs, figures, and direct observation using the newest instruments. The language of modern science, however, is missing from Kepler's dream narrative. Tycho Brahe, with his telescope and dense tables of astronomical observations, may exist at the periphery, but the *Somnium* remains resolutely populated with demons and witches. Yet, modern science exerts a considerable presence in the vast array of footnotes which threaten to engulf the text.

My analysis of the *Somnium* will proceed from the problems I have been discussing in my earlier chapters. These elements have not been chosen at random. As Ladina Bezzola Lambert notes in *Imagining the Unimaginable,* "three elements receive special prominence in the narrative of the Dream and are continually played off against each other: the book, the dream and the journey."[13] These elements seem very simple: they are essential objects of the fable, timeless evocations of a mythic time beyond time. They have figured prominently in my analysis up to this point. Taken individually or in combination, these elements do not have a single, stable, unchanging meaning. Instead, at a given period of time, each suggests a range of interpretive possibilities.

The book, the dream, and the journey are the most basic components of the *Somnium* as an allegorical text. Even the words, arranged next to each other in Lambert's phrase "the book, the dream and the journey" suggest a narrative, a procession from the one to the next. The book conceals the dream which, in turn, conceals the journey. The journey, so encased by the dream, helps to bring our attention to the dream as a space without which the journey is impossible. The dream within the book destabilizes the authority of the book. The conditions which make the dream valid are not the same as those which authorize the text. In this way, "the three elements mutually contradict each other: The physical journey to the Moon conflicts with the dream vision, and this in turn with the dream as book or literary dream. They can only exist next to each other through ambiguity."[14] And at the same time, each can stand in for the other. What is a book but an imaginary image transcribed onto paper? And the act of reading itself a metonymy for the physical journey of characters always as ephemeral as the apparitions of the dream?

My interest in the *Somnium* stems from its textual duality and apparent contradictions. This duality is most apparent in the contrast between the text, a *narratio fabulosa* clearly indebted to the cosmological allegories of twelfth century Neoplatonists, and the voluminous footnotes, so suggestive of the language that replaced and rendered irrelevant the proto-scientific musings of the Neoplatonists. Still, these two components of the text are not so clearly divisible from Kepler's own philosophical preoccupations. Instead, both parts of the text, if they can be viewed as separate, discrete units, reveal the influence of a philosophical and narrative tradition stemming from sources such as Plato and Macrobius.

THE BOOK

Kepler's *Somnium seu Astronomia Lunari* (*Dream or Astronomy of the Moon*) was originally written in 1609, but not published until 1634. At that time, Kepler was particularly interested by Plutarch's *De facie in orbe lunae,* a study of the face of the moon. However, the text had been started at a much earlier date. Kepler began work on his lunar astronomy as early as 1593, when he was still a student at Tübingen.[15] In this earlier project, the young Kepler was interested in demonstrating the validity of Copernicus' theory of heliocentrism. Under the tutelage of Michael Maestlin, a well-known astronomer of the time, Kepler became intrigued by the Copernican theory.[16] At this point, Kepler had not actually read Copernicus.[17] Instead, he knew of the theory from Maestlin's lectures. As this portion of a letter by Kepler demonstrates, he began to pursue his own proofs of Copernican theory at an early age:

> Already in Tübingen when I followed attentively the instruction of the famous Magister Michael Maestlin, I perceived how clumsy in many respects is the hitherto customary notion of the structure of the universe. Hence I was so very delighted by Copernicus, whom my teacher very often mentioned in his lectures, that I not only repeatedly advocated his views in the disputations of the candidates, but also made a

> careful disputation about the thesis that the first motion (the revolution of the heaven of the fixed stars) results from the rotation of the earth. I already set to work also to ascribe to the earth on physical, or, if one prefers, metaphysical, grounds the motion of the sun, as Copernicus does on mathematical grounds. For this purpose I have by degrees—partly out of Maestlin's lecture, partly out of myself—collected all the mathematical advantages which Copernicus has over Ptolemy.[18]

Kepler contrasts the "physical, or if one prefers, metaphysical, grounds" on which he bases his proof with Copernicus' mathematical approach to the problem. This contrast between the metaphysical and mathematical is striking to our modern sensibilities. The clumsiness of the "hitherto customary notion of the structure of the universe" perhaps stems from the errors of metaphysics applied to the natural world. Still, the apparent break between these two methods of proof is not so solid for Kepler.

The *Somnium,* while a fanciful allegory, is based on a rigid argumentative logic. However, the argument, by which I mean the allegorical narrative, by itself was not sufficient for Kepler. Between 1621 or 1622 and 1630, Kepler produced the footnotes, intended to provide an interpretive apparatus to a work "steeped in myth, legend and folklore as well as thinly veiled allusions to his own life."[19] Likewise, they are intended to exonerate Kepler's mother from the charges of witchcraft leveled against her due to an initially hostile response to a scientific text so clearly dominated by the supernatural.[20] Indeed, as Lawrence Lipking reminds us in "The Marginal Gloss," "Footnotes, as everyone knows, are defensive."[21] Though placed under arrest around 1615, Katherine Kepler "was actually imprisoned only in 1620, when she was taken to the prison in Leonberg. She was transferred to Gueglingen on August 25 of that year, and kept in the tower in solitary confinement, under unhealthy conditions, in heavy iron chains."[22] The variety of sources which Kepler conjoins both in the text and the notes are suggestive of the unique nature of the *Somnium.* They stress the extent to which the *Somnium* is built from a competing range of interests: classical allegorical sources, autobiography, and the scientific possibilities engendered by Copernicus. Further, even as Kepler attempts to detangle and articulate the meanings of the lunar voyage, he suggests even more elaborate levels of interpretation.

A "Geographical, or if you Prefer, Selenographical Appendix" follows the text and notes. This Appendix, written in 1623, consists of a letter written to Paul Guldin, a Jesuit mathematician. This letter is concerned, following Galileo, with observations of the "face" of the moon. Kepler notes the existence of circular craters on the moon. In keeping with his interest in the inhabitants of the moon, the Subvolvans and Privolvans of Levania, Kepler describes these craters as lunar fortresses. Their shape, to Kepler, resembles the walled city. His letter details the construction of such cities. Kepler's translation of Plutarch's *De facie in orbe lunae* completes the volume.

The *Somnium* opens on a dark and soon-to-be-stormy night. The narrator, presumably Kepler, falls into a deep sleep after watching the stars and the moon. These details of the narrative lend the exterior frame a kind of symmetry: the astronomer, after observing the stars, dreams of the cosmos. However, the narrator also recounts his interest in current political events. The first sentence of the *Somnium,* in fact, locates the night of Kepler's dream in 1608, a time significant for, the narrator tells us, "a heated quarrel between the Emperor Rudolph and his brother, the Archduke Mathias."[23] This introduction positions the *Somnium* within the actual events of Kepler's life. The historical context constitutes the outermost narrative frame. Kepler's time in Prague was relatively peaceful when compared to his residence in other cities. In Prague, for instance, Kepler was not persecuted because of his religious faith. Emperor Rudolf II was perhaps the explanation for this freedom from religious persecution. Rudolf II fostered the arts and sciences during his reign. By inviting Kepler and Brahe to his court, he ensured the position of Prague in the formation of the nova astronomia. Further, he had maintained peace, despite the conflicts ranging across Europe.

However, "the imperial throne did not radiate bright resplendence."[24] The Emperor's interest in art and science can be attributed, authoritative Kepler biographer Max Caspar tells us, to "a sick soul's capricious zeal for compiling."[25] The Emperor, "the strange, slightly built, unmarried eccentric on the throne"[26] was obsessed with his collections to an unhealthy extent. Thus, while the Emperor supported astronomy, "it is obvious that [. . .] state affairs must have suffered"[27] because of the Emperor's less than regal bearing. Indeed, Matthias seized control from Rudolph due to the Emperor's excessive interest in "wizards, alkymists, Kabbalists, and the like, sparing no expense to find all kinds of treasure, learn secrets, and use scandalous ways of harming his enemies."[28]

Religious opposition led to the creation, in 1608 and 1609 respectively, of the Protestant Union and the Catholic League.[29] The conflict between these groups, made up of the nobility, threatened the stability of the area, and led to the Thirty Years' War. As a result of escalating conflict, "as early as June 1608, Rudolph was forced to transfer the government in Austria, Hungary and Moravia to his brother Matthias," who had aligned himself with the Protestants.[30] The narrator indicates that he takes a great interest in the political controversy. He "turn[s] [his] attention to reading about Bohemia,"[31] presumably to greater understand current events. He mentions that, in his reading, he "came upon the story of the heroine Libussa, renowned for her skill in magic."[32] From there, the dream narrative actually begins. The reference to Libussa, and the narrator's interest in Bohemian history, seems an unlikely pairing until we consider both Kepler's and the Emperor's fixation on the supernatural. The narrator hopes to find in the past a message presaging the outcome of the current political situation. The essentially mystical character of the prophetic dream allows for this. We can perhaps attribute these features to Kepler's personal temperament, vacillating between the competing poles of astronomy and astrology; metaphysics and science; and Lutheran, Calvinist, and Catholic theological doctrines.

After falling asleep, the narrator finds himself at a market in Frankfurt. Here, he purchases a book detailing the life of Duracotus, an Icelandic astronomer. The story of Duracotus parallels Kepler's own, and is replete with persecution, mysticism, and astronomy. Like Kepler, Duracotus works with Tycho Brahe. In the narrative, Duracotus studies on Hven, a Danish island where Brahe maintained an observatory and headed an energetic and, at times, tumultuous household made up of family, servants, and students of astronomy. After his years of study under Brahe, Duracotus returns to Iceland. Upon returning home, he is reunited with his mother, Fiolxhilde, a wise woman equal to Brahe in her knowledge of the cosmos. Fiolxhilde has learned about astronomy the old fashioned way, however: she communicates with the daemons of Levania through magical ritual. Together, mother and son summon a daemon from the lunar tribes. Once invoked through a secret ritual, this daemon lectures on a variety of subjects, ranging from the geography and biology of the moon, the processes necessary to travel between Earth and the moon, to the contrasting perspectives of the universe provided by Levania and Earth. After this digression, the dream ends quite abruptly as a severe storm[33] breaks through Kepler's bedroom window.

THE DREAM

The narrative model employed in the *Somnium* can be viewed as a key factor contributing to the scientific significance of this text. Indeed, as Albert Schirrmeister points out in "Traum und Wissen in der Frühen Neuzeit," "Kepler weicht auf die literarische Form der Traumerzählung aus, um das anerkannte geozentrische Konzept anzugreifen" [Kepler yielded to the literary form of the dream narrative, out of which the geocentric concept was developed].[34] The textual history of the *Somnium* enhances Schirrmeister's interpretation. The text serves as a kernel which Kepler embellished with increasingly complex footnotes for many years. From this standpoint, the *Somnium,* as a single text, influenced and helped provide structure to Kepler's thoughts on planetary orbits. For Timothy Reiss, Kepler's nearly lifelong obsession with the *Somnium* testifies to its central importance among his works.[35] In other words, this was not an unimportant or frivolous exercise that Kepler engaged in to take his mind off of the dire political situation of his time. Instead, the *Somnium,* as text, provides an imaginary space where Kepler can examine the possibilities of the new astronomy.

Like the epigenesis of a scientific idea or project itself, the embedded, narratologically ornate, hermetic structure of the *Somnium* as allegory traces the growth in knowledge of a whole career: at the most practical level, Kepler added to the footnotes throughout his life, embellishing a text he began as a student; Kepler's use of the dream narrative as a generic model for the *Somnium* is significant in that this particular narrative model intersects several significant areas of concern for medieval and early modern thinkers. While Schirrmeister points out that the *Somnium* leads to the theory of elliptical orbits, I would argue that the narrative model of the *Somnium,* with its attendant theological, linguistic, and philosophical concerns, is perhaps more important in its contribution to Kepler's theory of elliptical orbits as articulated in the footnotes of the *Somnium.* The specific text adheres to the key factors so necessary for the narrative model of the cosmological dream allegory. The narrative model, in its most basic skeletal form, shapes Kepler's scientific ideas in ways clearly underestimated by science historians who relegate the *Somnium* to the status of an insignificant, merely curious, document. The *Somnium* is a specific text that belongs to a finite genre. Thus, Kepler's *Somnium* can be classified as a cosmological dream allegory, but our classification can not stop with a mere naming of the genre to which the text belongs.

As a cosmological dream allegory, the *Somnium* contains an exterior narrative frame featuring a sleeping dreamer and an account of a cosmic journey. These two necessary factors, however, compete to retain their relevance in the face of the footnotes they engender.[36] The combination of these two factors, their coexistence in a single narrative, likewise hints at a series of concepts or thematic preoccupations which will influence or guide the narrative. In other words, the most basic components of plot hint at the corresponding usefulness or desirability of the model. This narrative model consists of two basic actions: sleeping and traveling. Furthermore, it consists of at least two narrative frames. We can conceive of these frames as textual parallels to the interlocking spheres which constitute the Neoplatonic model of the universe. I note this because the Neoplatonic model of the universe is not itself without consequence for determining the structure of this type of narrative. Kepler allegorizes, in short, narratological structures as astrophysical realities and vice versa.

This model, built as it is from correspondences between the macrocosmos and microcosmos, itself greatly influences the characters of the narrative: the verifiability of the dream is influenced by the positions of these spheres at a particular time. Kepler, while working towards a theory which contributed to disproving such theories of correspondence, was, recall, a writer of horoscopes.[37] The horoscope is itself a genre dependent on such a dynamic of correspondence. Indeed, his long resistance to recognizing the infinity of the universe and his focus on the solar system as a cohesive entity testify to his personal attachment to the philosophical concepts underlying a Neoplatonist model of the universe. As J.V. Field remarks in *Kepler's Geometrical Cosmology,* "Kepler stands out as a supporter of the older ideas, at least to the extent of believing that the Solar system was a uniquely important component of the Universe."[38]

The cosmological dream allegory, Kepler's included, relies on this principle of correspondence. However, the concept becomes of greater importance when we consider the impact of Kepler's *Somnium* on the significatory possibilities available to the cosmological dream journey as an allegorical motif. The dual role of Kepler the scientist and Kepler the mystic has been frequently commented upon. His own writings bear testament to this heavily divided character. Perhaps the appeal of Kepler as a figure in the history of science stems from his adherence to viewpoints which, from a modern standpoint, seem irreconcilable. We can, however, understand that, for Kepler, astronomy was an act of theological devotion. Indeed, it is possible to view Kepler's *Somnium,* as Albert Schirrmeister remarks, "in einen religiös prophetischen Kontext" [in a religiously prophetic context].[39] Thus, correspondences between celestial bodies and earthly realities determine the range of philosophical and theological concepts evoked by the motif of the dream of the cosmos for Kepler and his predecessors.

It is not by chance that the text so clearly matches the features of the narrative template I offer of the cosmological dream allegory. In note 2 of the *Somnium,* Kepler acknowledges the direct influence of the texts which employ a similar structure. In documenting his selection of Iceland as the setting, he notes that "in this remote island I perceived a place where I might fall asleep and dream, in imitation of the philosophers in this branch of literature."[40] With this statement, Kepler establishes a double displacement of place and conscious state. Both the island and the dream are necessary for

Kepler to successfully imitate the philosophers. Kepler goes on to list these influential sources, and to acknowledge their pairing of journey and dream:

> Cicero crossed over into Africa when he was getting ready to dream. Moreover, in the same western ocean Plato fashioned Atlantis, whence he summoned imaginary aids to military valor. Finally, Plutarch, too, in his little book on *The Face of the Moon* [. . .] ventures out into the American ocean [. . . .] [Lucian] too, [in his *True Story*] sails out past the Pillars of Hercules into the ocean and, carried aloft with his ship by whirlwinds, is transported to the moon.[41]

Not all of these texts feature the dream, however. Plato attributes the story of Atlantis to an Egyptian priest. Likewise, Lucian does not mention a dream in his story. On the other hand, both of these sources also suggest the necessary displacement of conscious states achieved otherwise through the narrative frame of the dream. The tale recounted by the priest suggests the possibility of a mystic state. Atlantis, already a setting evocative of the supernatural and otherworldly realm of the dream, is recounted by an Egyptian priest who, perhaps, received his vision of the island in the spiritual darkness produced by the serpent Apep as it engulfs the sun. Lucian likewise achieves this displacement of consciousness simply by passing through the Pillars of Hercules. Geographically, these merely denote the rocky borders of the Straits of Gibraltar. However, in a mythographic context, these connote the boundaries of logic and the knowable. Allegorically, Dante's evocation of these boundaries

> insinuates the oblique and shadowy path of metaphoric language in which truth and fiction have a simultaneous existence and the presumed unity of sign and meaning is shattered. He decidedly obliterates, in other words, the distinction between allegory of poets and allegory of theologians conventionally based on the fictive or nonfictive status of the literal sense.[42]

Kepler, like Plutarch and Lucian, passes through a similar gate. The dream itself emanates from Somnus' twin gates of pearl and horn. Duracotus and Fiolxhilde summon the Daemon while protected by a magical circle, itself a gate allowing them to access, but not inhabit, the Daemon's cryptic realm. But these gates are not merely dividers or partitions. Though Kepler describes the moon as an *insula,* it differs from Plutarch's allegorized island. The ascent to the moon impels complexity, polysemy, in a way that a water voyage, no matter how much the elements buffet the ship, cannot. In the following three chapters, I propose that the *Somnium,* far from a historical footnote, enables the allegorical motif of the cosmic voyage in a new way. However, I maintain that the full range of meanings available to this motif cannot be separated from the Neoplatonic cosmologies which preceded the *nova astronomia.*

Chapter Seven

The Poetic Structure of the Circle

The astronomer, ensconced in a chalk circle, measures the orbits of circling planets, summoning their power. The perfection of the circle impels perfectible motion; the mystical perfection of its shape protects the magician from monstrous forces eager to consume unwary wizards. Or does it? The necromantic astronomer, despite his fastidiously fashioned magic circle, notes considerable discrepancies between the ideal circular orbits and the actual movements traced by Mars and Volva. How can this magic circle, if not a microcosmic representation of perfect cosmic proportions, protect our magi-astronomer in a universe already maligned by unholy forces?

THE MAGIC CIRCLE

As I argue in "Allegory and Movement," the philosophical and theological implications of geometric shapes provide the basis for pre-modern astronomy. But how does the circle, that mark of singularity and eternity whether traced in chalk or star dust, figure into an astronomy faced with imperfect forms?[1] In this chapter, I will examine this question in relation to a similar discrepancy between the supernatural and scientific evidenced in Kepler's supposed "creation" of science fiction as a genre. In other words, how do the modal choices of a scientist as a writer reflect his philosophical and theological preoccupations?

The heavy emphasis on metaphysics and theology evident in Kepler's scientific writings results in, as Gerald Holton affirms, a multimodal approach reliant on intuition and inspiration:

> With rich imagination he frequently finds analogies from every phase of life, exalted or commonplace. He is apt to interrupt his scientific thoughts, either with exhortations to the reader to follow a little longer through the almost unreadable account, or with trivial side issues and

> textual quibbling, or with personal anecdotes or delighted explanations about some new geometrical relation, a numerological or musical analogy. And sometimes he breaks into poetry or a prayer—indulging, as he puts it, in his 'sacred ecstasy.'[2]

Holton's account of Kepler's writing appears to stigmatize Kepler's preoccupations. He notes that, "when [Kepler's] physics fail, his metaphysics come to the rescue."[3] Kepler's 'sacred ecstasy,' then, clearly detracts from Kepler's 'scientific thoughts.' Indeed, this passage from Holton generically marks each type of writing and also assigns each a value based on empirical claims to truth. Thus, he interrupts his more valuable scientific thoughts with "trivial side issues." I bring up this example, however, because much of the *Somnium* so clearly unsettles our own modern valorization of issues, an observation made by many who have studied this work.

In *Voyages to the Moon,* for example, Marjorie Hope Nicolson provides a reading that does not maintain the separation of modes that Holton so rigidly upholds. For Nicolson, not only is Kepler's writing multimodal, but also, in the case of the *Somnium,* multigeneric, a fusion of the supernatural and scientific: "no important later voyage will employ so fully the supernatural, yet none will be more truly 'scientific' than the Dream, which was the *fons et origo* of the new genre, a chief source of cosmic voyages for three centuries."[4] Roger Bozzetto makes a similar claim in "Kepler, naissance de la visée spéculative fondée sur la science au sens moderne du terme." He notes that,

> il me paraît le premier représentant de ce qu'est un 'primitif' de la SF, et même de la 'hard science fiction.' Il constitue ce 'chaînon manquant' entre les textes d'imagination pure de Lucien et les aventures appuyées sur les découvertes scientifiques d'un Cyrano de Bergerac, puis d'un Jules Verne. C'est en tout cas l'une des premières fictions conjecturales, ou spéculatives.
>
> it seems to me the first representative of that which is a primitive science fiction and also hard science fiction. It constitutes a 'missing link' between the purely imaginative texts of Lucian and the adventures founded on the scientific discoveries of a Cyrano de Bergerac, then a Jules Verne. It is in any case one of the first conjectural or speculative fictions.[5]

Bozetto identifies the *Somnium* as an intermediary point in a continuum stretching from the mythically informed texts of writers like Lucian and Plutarch to the scientifically-informed fictions of later writers such as Jules

Verne. Nicolson, in accord with Bozzetto's argument, claims that, besides merely marking a transition, the *Somnium* completely exemplifies the qualities of the genres which it surpassed while providing a blueprint for a new literary genre.

In *The Sleepwalkers: A History of Man's Changing Vision of the Universe,* Arthur Koestler likewise applauds the *Somnium* as "the first work of science-fiction in the modern sense."[6] Such assertions regarding literary genre, however, make some wide-sweeping claims that we can not ignore. For example, to what extent does the *Somnium* qualify as modern science fiction or the first example of speculative fiction if its author remains so resolutely medieval? In other words, such claims may lead us to question how the "unreadable," mystically inclined Kepler produced a genre so indicative of the modern, the rational, the scientific. We can find an answer, or at least the suggestion of an answer, to this paradox of genre by examining changing attitudes toward circular motion, a subject equally embroiled in a morass of the mystical and empirical.

The poetic structure of the circle, then, impels the following question: How do philosophy and empirical evidence affect narrative genre? The argument I explore endows the fictional frame with a power like that of the magic circle. Not merely a diversionary tactic or "palliative add-on,"[7] the narrative belongs to and creates a new literary genre. Kepler's creation of this genre, correspondingly, coincides with a host of philosophical and theological concerns worked out and added to over a long period of time. To this extent, the genre of the *Somnium* emerges over a lifetime, marking not just a turning point, but an entire period of transition. The hybrid genre that Nicolson and Bozzetto claim for this work, then, must also reflect shifting attitudes toward a variety of philosophical and theological issues, such as the emergence of the circle as a purely geometric figure.

The combination of genres, I contend, results in a new genre which fulfills and surpasses the earlier categories. We could distinguish these genres through the use of markers such as "imaginative" and "speculative," but in doing this, we must understand which narrative features allow us to make these distinctions. My emphasis on motion and the means of conveyance depicted in narratives provides us with a way to distinguish these genres in terms of the vehicles employed in each. Thus, the imaginative fiction may employ a dream journey or some kind of supernatural mode of transport. As in the *House of Fame,* such a narrator may find himself snatched into the night sky. Or, as in Alain de Lille's *Anticlaudianus,* such a narrative may present a vehicle built by allegorical personifications. The work of speculative fiction, on the other hand, features a technologically-driven cosmic voyage, or shows the self-reliant narrator constructing a means of astral ascent. From a modern perspective, these two missions, and the vehicles they presuppose, appear incompatible. The pragmatic scientist seldom undertakes the mystical journey. The shaman soars through the cosmos, unaided by Black and Decker. So, Bozetto and Nicolson's statements seem difficult to fully support: how can a text be completely supernatural and scientific at the same time?

THE FAUSTIAN DEBT

Despite Kepler's deep spirituality, he bears more than passing similarity to Faust. The legendary Faust, as a product of Kepler's own era, writes with and is written by the magical figures of hermetic and alchemical knowledge. Goethe's *Faust* provides an example of this connection between metaphysical quandry and the search for macrocosmological knowledge which propelled much of Kepler's own writing. Kepler and Goethe were not, of course, contemporaries.[8] But Kepler and Goethe's Faust could have been contemporaries, despite the massive differences at the root of their respective quests. Indeed, Kepler's link to Paracelsus, "the most famous representative of the occult philosophy of the Renaissance [. . .][and] one of the prototypes for the composite character of Dr. Faustus,"[9] attests to this similarity. At the end of *De stella nova,* for instance, Kepler heralds a new era revitalized by the achievements of both Copernicus and Paracelsus. But the concerns of Kepler and Paracelsus differ from those of Goethe's Faust. Most significantly, while one can argue for medieval conceptions of the sublime, the theological and philosophical concepts underlying this sublime differ markedly from those of the more familiar concept paired with Romanticism.

Much of Act I of Goethe's play details Faust's own immersion in the sciences. In the Nacht section at the beginning of Act I, Goethe depicts the various stages of Faust's attempts to find and experience the Transcendental. While we can assume that the narrative compresses this process, Faust quickly confronts a series of types and modes of knowledge. In fact, the tragedy begins with Faust's famous declaration of his learning and its limits:

> Habe nun, ach! Philosophie,
> Juristerei und Medizin
> Und leider auch Theologie—
> Durchaus studiert, mit heißem Bemühn.
> I have, alas, studied philosophy,
> Jurisprudence and medicine, too,

And, worst of all, theology
With keen endeavor, through and through.[10]

He clearly expresses his own exasperation of the ultimate emptiness at the root of this learning. Despite his mastery of both Trivium and Quadrivium, Faust endures a deepening and unassailable sense of meaninglessness. Perhaps it would be better to phrase the previous statement in this manner: *because* of his mastery of both Trivium and Quadrivium, Faust endures this purposelessness. Of course, Theology, more so than Philosophy, Jurisprudence, or Medicine, has led Faust to the declaration of his quest, of his desire to move beyond knowledge and toward pure experience. On the one hand, scholarship has traditionally, and rightly, aligned Faust's own indictment of knowledge with the overwhelming aesthetic and philosophical program of Romanticism. Faust's indictment of Theology, for the Romantic program, liberates the aesthetic imagination from the confines of official Catholic doctrine in the same way that citizens liberated themselves from the tyranny of monarchy.

However, viewing *Faust* as a justification of, or precursor to, Romanticism as it developed in the nineteenth century, we would view Theology as clearly distinguishable from Philosophy, Jurisprudence, and Medicine. The irony of Faust's statement that, of all of these disciplines, Theology is somehow the worst of all underscores the extent to which Goethe's *Faust,* as an articulation of the Romantic program, both is and is not concerned with the spiritual.

For Goethe, these different branches of learning are quite distinct, but for his character Faust's historic precursor, they are all part of a tradition of learning that stems from Theology. Thus, Jurisprudence, Philosophy, and Medicine are likewise of no assistance in Faust's quest for absolute knowledge. Theology is not condemned because of its uselessness. Instead, the "real" Faust, spinning tales of diabolism and telling fortunes in tiny towns sprinkled across the Bavarian countryside, despairs because of the central role Theology has had in the formation of these other areas of knowledge.

Goethe's division of these areas of learning reflects their value in relation to Romanticism. To this extent, theology is the least significant, the most useless, the most emblematic of what needed to be dismantled. However, if we think of Faust, like Kepler, as a cosmologist interested in untangling the actual order of the universe from how it was conceived in strictly theological terms, then an entirely different problem appears. The other disciplines are also of no relevance to Faust because they cannot be separated from the theological concerns. Instead, even if separated from theology, they are little more than practical applications of the covenant between God and man, or of moral problems. Even medicine, to modern readers the most readily scientifiable of these disciplines, was tied quite closely to moral and ethical questions. Furthermore, the body and the universe were, as Goethe's Faust indicates when he declares "Ich Ebenbild der Gottheit,"[I, image of the divine][11] still unimaginable outside of Theology.

Despite Kepler's reliance on theology and his mysticism, his writing, as Holton notes, "prob[es] for the firm ground on which our science could later build."[12] Holton does not argue, however, that Kepler transcends some sort of pre- and post- scientific boundary. Otherwise, we could identify and isolate the non-scientific qualities of Kepler's writing as idiosyncrasies of an earlier period. But Holton states that the modern impression of Kepler's textual preoccupations is more than "the result of the inevitable astigmatism of our historical hindsight."[13] Instead, he indicates that the basis for a modern science, devoid of mysticism, can be found in the "ascetic standards of presentation originating in Euclid."[14] Further, Galileo, Kepler's contemporary, does not reveal the same interest in the supernatural so clearly evident not only in the *Somnium,* but in all of Kepler's writings.

Holton points out that Kepler's writing process draws Kepler's spiritual preoccupations to the foreground. Thus, while Faust despairs because of the central role of Theology, Kepler revels in its centrality. For early modern scientists like Kepler, the question of scientism versus mysticism involved the mode of presentation. The modern method of science writing, nascent in Euclid and distinguishable as a separate mode in Newton's *Principia,* is a genre in the process of becoming.[15]

However, the orderly mode of presentation expected by scientific writing does not always arrive at the empirical truth it seeks and models itself upon. Kepler's reliance on the imaginative, while lending his writing an air of mysticism, also moves him away from an idealized model of order valorized by geometry. Indeed, his initial pairing of physics and astronomy, to modern readers an essential pairing, was arrived at through his intuitive sense that these belonged together, despite the objections of his contemporaries.

THE HAUNTING SPELL OF CIRCULARITY

The order of presentation presaged by Euclid relies on concealment by the scientist/ writer. The scientist hides or elides the random elements that appear to play no major role in the solution of a problem. Such a strategy assumes an ideal or a set of conditions which determine relevance. One well-known and frequently commented on instance of this discrepancy comes from Galileo's inability or lack of desire to recognize the validity of Kepler's theory of elliptical orbits. The boundaries between scientific and supernatural are not as clearly identifiable as may be surmised. In this case, Galileo's pursuit of order itself reveals an attachment to the aesthetic, and not empirical, criteria governing astronomical observation. John Herman Randall's "The Development of Scientific Method in the School of Padua" was one of the first essays to argue this point. While this essay was published in 1940, its influence has not reformed the customary viewpoint of Galileo as a kind of mythic hero of empiricism triumphing over a corrupt mysticism. To this extent, as W. Roy Laird argues, "Galileo is still often portrayed as having overthrown Aristotelian natural philosophy and method in order to found his new science of motion."[16] The collection *Method and Order in Renaissance Philosophy of Nature: The Aristotle Commentary Tradition* actively contests this view, presenting Galileo not as the reasoned and logical counterpoint to Kepler's mysticism, but as a figure likewise impacted by "the remnants of the medieval tradition."[17] Discussion regarding this medieval tradition tends to focus on circularity, a conception closely linked to belief in the perfection of cosmic design. To Galileo's credit, Kepler, though "one of the first astronomers to be free of perhaps the most objective basis for circularity, the assumption of solid spheres," also "hesitated long before abandoning the circle."[18]

Alexandre Koyré, like Pierre Duhem an early advocate of the poetic structure of scientific texts, uses the phrase "haunting spell of circularity" to describe Galileo's adherence to a model of the universe that did not correspond to the most exact physical data available on the subject. Resolving the irregular motions of the planets remained a key goal of astronomers until Kepler's first planetary law.[19] This argument, as has been explored by, among others, Koyré, Panofsky, and Hallyn, is not confined to the realm of physical observations, however. Indeed, as Erwin Panofsky comments in *Galileo as a Critic of the Arts,* "at the very beginning of the *Dialogue,* Galileo unequivocally endorses the belief [. . .] in the perfection [. . .] of the circle not only from a mathematical or aesthetic but also from a mechanical point of view."[20] Celestial bodies, for Galileo, move along circular paths because of the perfection of the circle as form and the universe as an expression of Divine perfection. Kepler, on the other hand, abandoned this idea based on the observations he had culled from Brahe. No circle could match these points, and Kepler "had been unable to understand how it was physically possible for a planet to follow an eccentric circular path."[21]

However, this breaking point between the two great astronomers has also been theorized along aesthetic guidelines. Koyré's argument operates on a division between classicism and mannerism. The perfection of the circle is closely linked to classicism, both philosophically, through the evocation of Plato's forms, and aesthetically, through the adherence to perspective so assiduously pursued by Renaissance painters. Artistic theories of pictorial perspective, based on the principal forms, also located these forms in nature. As Hallyn reminds us, "Renaissance art generally sought to unify the representation of the human body in terms of the circle."[22] Indeed, Girard Desargues' development of projective geometry in the *Brouillon project* (1639), which was derived from Alberti's treatment of perspective in *De pictura* and first systematized the process by which different shapes share the same origin, was ignored until the nineteenth century.[23]

The convergence of problems of artistic representation and geometric computation surfaced in numerous disciplines. As Hallyn notes in "Du Monde de Kepler Comme Anamorphose," "mais surtout [. . .] le mode de construction du cosmos géocentrique répond peut-être le mieux à celui qui est inhérent à la perspective géométrique de la Renaissance" [but above all [. . .] the mode of construction of the geocentric cosmos responds best to that which is inherent in the geometric perspective of the Renaissance].[24] For one, the architectural symbolism of the cathedrals conveys correspondence between the Divine order of God's kingdom and basic geometrical forms. Furthermore, the omnipresence of problems of perspective likewise extended to living beings, so that "a kind of grammar was [. . .] constructed, according to which the parts of the body were subject, as they completed the most varied movements, to circular units of measurement."[25] The meditation on the perfection of the circle was not limited, then, to the circle as a static form. Instead, this form, and the motion it impelled, expressed and were subject to the ineffable rules of divine cause.

Galileo conceived of circular motion as the ideal form of movement. In this way, he reflects the influence of an Aristotelianism that "still represented a more comprehensive and internally coherent system than any that was available to replace it."[26] Hence, his rejection of Kepler's ellipses extends beyond an aesthetic disinclination towards Kepler's contorted orbits. These shapes, besides not conforming to the geometrical order of the circle, likewise describe movement that does not follow from a classical conception of celestial harmony. These differences have been used to show Kepler's celestial mechanics as aesthetically linked to mannerism. The arguments for this are quite convincing. However, I think that, as a result of these arguments, we are too compelled to view Galileo and Kepler as exemplars of the Renaissance and Baroque, thereby reinforcing the opposition between them. The tendency is too strong to see each as fully representing the aesthetics of the Renaissance or the aesthetics of the Baroque in their viewpoints. While this may be the case, my examination of these arguments leads me to the influence of Neoplatonic models of the cosmos in each astronomer. The shape of this influence differs for Galileo and Kepler. While Galileo and Kepler's astronomical theories demonstrate the influence of Renaissance and Baroque aesthetics, their formulations also stem from the Neoplatonist models of the universe that preceded the *nova astronomia.* This distinction between Galileo and Kepler reveals the extent to which the Neoplatonist influence asserts itself in the aesthetic criteria each figure attributes to the cosmos.

The exact combination or source of these influences affects the direction of their work. Thus, Kepler and Galileo embody the two dominant themes of

> the scientific revolution of the 17th century—the Platonic-Pythagorean tradition, which looked on nature in geometric terms, convinced that the cosmos was constructed according to the principles of mathematical order, and the mechanical philosophy, which conceived of nature as a huge machine and sought to explain the hidden mechanisms behind phenomena.[27]

Indeed, Kepler never abandoned his faith in the geometric model of the universe, exemplified by the nested solids of the *Mysterium Cosmographicum,* indicating that, though "[t]his idea was not in the mainstream of Christian thought," "[f]or Kepler God is first and foremost a geometer."[28] But the customary argument that an outmoded mysticism influenced Kepler while Galileo served as exemplar of a clear new logic can't be readily accepted. Galileo's attitude toward the circle attests to the complexity of this influence.

Galileo rejected, or refused to acknowledge, Kepler's ellipses on the basis of his aesthetic valorization of the circle; this valorization of the circle already has its roots in Neoplatonist thought. The circle was a perfect form that, despite its perfection, was not transparent. It did not reveal all of its mysteries. Thus, as an ideal geometric form, the circle also symbolized the limits of human knowledge: "of all the truths of geometry, two, its beginning and its end, as Dante had said in the *Convivio,* cannot be demonstrated: the point, which is not measurable, and the circle, which cannot be squared."[29] The geometer, proceeding by demonstration, still encounters problems which cannot be resolved through such demonstration. The squaring of the circle, as evoked by Dante, "constitute[s] the paradigm of the impossibility of man's understanding everything."[30]

The problem of the squared circle was one to be solved through geometric methods. It is a problem that cannot be solved by geometric methods, however; this problem is only describable or identifiable through recourse to these same methods. While geometry is based on tangible proofs, it also suggests the limits of human endeavor. In Canto XXXIII of the *Paradiso,* Dante is struck by his inability to comprehend and articulate what he sees in the cosmos. He likens his efforts to those of the geometer:

> Qual è 'l geomètra che tutto s'affige
> per misurar lo cerchio, e non ritrova,
> pensando, quel principio ond'elli indige,
> tal era io a quella vista nova:
> veder voleva come si convenne
> l'imago al cerchio e come vi s'indova
> ma non eran da ciò le proprie penne.

> As is the geometer who wholly applies himself to measure the circle, and finds not, in pondering, the principle of which he is in need, such was I at that new sight. I wished to see how the image conformed to the circle and how it has its place therein; but my own wings were not sufficient for that.[31]

For Dante, geometry, through its rigorous method of demonstration, serves as the ideal motif for his own struggle with representation. Though the geometer proceeds through proofs, this logical science still includes problems that are impossible within its logical framework. The same problem applies to writing, which provides a means of "measuring" phenomenon, but can also surpass its own limits.

We cannot forget that, for the early modern imagination, geometry was not an activity separated from even Biblical allegoresis, serving as a method for interpreting and articulating the shape of the book of the world. Allegoresis, or the method of disentangling hidden meanings, influenced "philosophical schools [as well as] history and natural science."[32] The squaring of the circle provides an example of an unsolvable equation grounded in a tradition that held that "the gods express themselves in cryptic form—in oracles, in mysteries."[33] This classical notion remained in medieval thought, evident in Christian writers such as Augustine, and contributes to the Neoplatonist concept of the world as text.

THE SHAPE OF THE COSMOS

Kepler's victory in determining the shape of the cosmos testifies to an ability to unravel Divine mystery suggestive of something greater than puzzle solving. Instead, the solution of such a problem hints at the ability to create on the level of Divinity. Kepler thus synthesizes Copernicus and Ptolemy's views towards the relation between humanity and the divinity of the cosmos. Copernicus based his *De revolutionibus* on the structure of Ptolemy's *Almagest.* His attitude toward this important earlier source suggests that he used it as a revered model. However, there is a discrepancy in Copernicus and Ptolemy's attitudes towards the ability to accurately determine the shape of the cosmos. Ptolemy recognized the fallibility of even the most logical of methods for making astronomical calculations: "But the author of the *Almagest* counterbalanced praise of astronomy with consciousness of the fundamental uncertainty of all mathematical representations."[34] Copernicus, on the other hand, sees the ability to solve mathematical problems as a direct result of the Divine nature of humanity. Thus, in contrast to Ptolemy, Copernicus declares that "it is highly unlikely that anyone lacking the requisite knowledge of the sun, moon, and other heavenly bodies can become and be called godlike."[35]

Hallyn summarizes the main distinction between Ptolemy and Copernicus on the relation between Divine and human as follows: "In place of a universe whose beauty and rationality escape us, and which thereby calls us to humility, Copernicus substitutes a cosmos for which man is the final purpose and whose true plan he can reconstruct."[36] In such a formulation, the power and value of knowledge centralizes the position of humanity in the cosmos. Hallyn likens Copernicus' view towards astronomy with Marsilio Ficino's *Theologica platonica,* which associates knowledge of the cosmos with creation of the cosmos:

> And so, since man has seen the order in the heavens, its progressions and proportions or results, how could anyone deny that he possesses almost the same genius as the author of the heavens and that he could, in a manner of speaking, create the heavens, if he found the instruments and the celestial matter, since he creates them now, in another manner of speaking to be sure, but according to a similar plan.[37]

Kepler embodies qualities which seem contradictory as they are portrayed in the writings of Ptolemy and Copernicus, and views the reconstruction of the universe as, itself, an act of the most pietistic humility. His synthesis of these qualities is telling of the variety of sources and styles so evident in his writing. Lutheran and Calvinist theology affirm Kepler's feeling of humility before the shape of the universe.

Copernicus' view, on the other hand, suggests that human endeavors ensure Divinity. The quality of Divinity is not innate for Copernicus.

Instead, it is arrived at through "the requisite knowledge of the sun, moon, and other heavenly bodies." In this sense, his approach differs considerably from that suggested by patristic writers. In his *Enchiridion,* for example, Augustine urges Christians to acknowledge that "'It is enough for the Christian to believe that the only cause of all created things, whether heavenly or earthly, whether visible or invisible, is the goodness of the Creator.'"[38] Still, the competing Biblical and classical sources evident in the Scholasticism that prompted the new astronomy required "an unbounded faith in the power of human reason to solve the problems of nature."[39]

Kepler, like Copernicus, uses the image of God as an engineer, a Divine Clockmaker, to resolve competing theological and scientific claims. This image, derived from *Questions on the Eight Books of Aristotle's Physics* by Jean Buridan, suggested that "the heavens and the earth were at least tentatively subjected to a single set of laws."[40] Furthermore, this conception of "the heavens as a celestial mechanism, a piece of clockwork, [. . .] break[s] the absolute dichotomy between the superlunary and sublunary regions."[41] Copernicus' references in *De revolutionibus* to the Creator as *opifex,* as the "Optimi opificis," or Most Excellent artisan,[42] while not new, were brought to the foreground in the Renaissance "as part of the reaction against scholasticism, and Copernicus extended its implications for the scientific enterprise."[43]

Kepler, like Copernicus, relied on the mechanical metaphor, above all, as a way to explain celestial movements. In the *Nova Astronomia,* he sets out:

> 'to show that the heavenly machine is not a kind of divine living being but similar to a clockwork in so far as almost all the manifold motions are taken care of by one single absolutely simple magnetic bodily force, as in a clockwork all motion is taken care of by a simple weight.'[44]

Kepler does not deny or circumvent the divinity of the cosmos, but instead indicates that this divinity has produced a universe which operates like a clockwork, or which has been set in motion by a process of movement. The divine is not innate in movement: God does not produce the movement of the planets. Instead, God designs the process by which the planets move. For Holton, Kepler's theory of motion demonstrates the influence of "the Lutheran God, revealed to him directly in the words of the Bible" and "the Pythagorean God, embodied in the immediacy of observable nature and in the mathematical harmonies of the solar system."[45] This concept, stemming from impetus dynamics, or the idea that a projector imposes force onto a moving body, was used to refute Aristotelian conceptions of motion and was "extended [. . .] from the earth to the heavens."[46]

The combination of views accompanying the formulation of a pre-Newtonian theory of motion stresses the varying degrees of theological influence in the thought of early scientists. Indeed, as we have seen, the theory of motion divides Galileo and Kepler. This combination results from the order of science at that time, which "should not be thought of as being single and well-defined" and should, instead, be conceived of as "a nesting and entwining of several different orders, some of which remain static over time and others that are in a state of transformation and change."[47] This conflict, then, ultimately reflects something more than two scientists' estimation of a physical concept.

Instead, their differing views on an interpretation of the Aristotelian concept of motion extend to the method of presentation favored by each. As Edward Rosen points out, at least part of Galileo's reluctance to accept Kepler's theory of celestial mechanics stems from Kepler's interpretation and presentation of such philosophical concepts as the principle of motion.[48] Thus, "Kepler's obscurity, prolixity and mysticism [. . .] were so repugnant to Galileo that he had no desire to seek out 'the nuggets of real gold hidden away in Kepler's heap of dross.'"[49] To this extent, the perfection of the circle as a figure arises from a theory or conception of motion. Figures such as the circle and the ellipse describe motion, a process invoked, for example, by Edmund in *King Lear* as he laments the fatal influence of "spherical predominance" on his own social position, remarking on the negative effects of the "enforced obedience of planetary influence."[50]

Likewise, the depiction of motion itself in narrative, or as a constitutive component of an allegorical text, resounds with philosophical and theological implications. Thus, the plot of the *Somnium* relies on the narrativization of motion. The elements of the narrative that have earned it distinction as the first recognizably modern science fiction narrative reveal the very non-modern discourse of motion as a theological concept. As the narrative begins, we learn that Duracotus' mother Fiolxhilde "often used to take me up to the lower slopes of Mt. Hekla,"[51] a place where, we are told, those who transcend the laws made by "dull minds" may perish in the mountain's volcanic chasms. Here, Kepler likens these journeys to Hekla with the story of Empedocles, who "hurled himself into the crater" of Mt. Etna and "sacrificed himself alive."[52] At this point in the notes, Kepler provides a list of legends of those who have made journeys to sacrifice themselves for knowledge, referencing not only Empedocles, but also Gaius Pliny, Homer, and even Aristotle. The desire for knowledge impels these quests and, in the stories Kepler recounts, the movement of each character allegorizes the process of finding knowledge.

On Hekla, Duracotus and his mother seek herbs, which Fiolxhilde uses for their magical properties. She "made little bags out of goatskin, which she filled and carried to a nearby port to sell to the ships captains;" when Duracotus opens one of these magical wind-summoning bags, she "made [Duracotus], instead of the bag, the property of the skipper."[53] At this point, the initial quest for forbidden knowledge, depicted as a hike in the mountains, becomes the necessary transit of Duracotus, his name itself intended to suggest an island or a place on the edge of the sea,[54] from one island, Iceland, to the other, Hven.

Further, Kepler does more than depict various journeys at the beginning of the narrative. He also identifies Duracotus with a magical means of transport, showing Fiolxhilde's substitution of her magical items for Duracotus. In this way, Duracotus undertakes journeys and emblematizes the journey of discovery. Fiolxhilde exchanges Duracotus for the bag after he opens the bag and discovers its contents: "herbs and [a] linen cloth embroidered with various symbols."[55] After she sells Duracotus, and not the magical bag, to the skipper, the ship departs "unexpectedly," the following day, presumably owing to the "favorable wind" that Duracotus mentions.

But Duracotus is not the only character who may allegorically represent motion itself. In fact, the apparently bodiless Daemon who teaches so much to Duracotus and his mother in the form of a mysterious voice suggests that the "very busy" daemons of the moon are, in fact, the shadows created by eclipses. Like Duracotus, his name an allegorical figuration of an island, his life mapped as a journey between islands, the Daemon inhabits another island, Levania, or the moon. In describing the transit from the moon to earth, for example, the Daemon tells us that when a shadow cast by the moon (because of the position of the sun) "touches the earth with its apex [. . .] we rush toward the earth with our allied forces."[56] This earthly invasion by daemonic forces, however, is only "permitted [. . .] when mankind sees the sun in eclipse," a point reiterated several times as the Daemon describes not only the transit of daemons to earth, but also the movement of humans to the moon. Significantly, only humans "most devoted to" the daemons of the moon are permitted transit; the narrative indicates here that this group of daemonic devotees includes "dried-up old women,"[57] those who "choose to spend their time in the constant practice of horsemanship or often sail to the Indies,"[58] and "philosophers who zealously cultivate all the philosophical sciences (namely, the family of these spirits)."[59] And, while Kepler likens these spirits to the philosophical sciences, he also tells us that evil spirits, "called powers of darkness and of air," should only naturally be "banished to the shadowy regions, to the cone of the earth's shadow."[60]

In the notes that explain this complex set of associations, Kepler tells us that the daemons also inhabit shadows because shadow measurement (or sciametry) reveals knowledge of celestial phenomena. But he also continues to liken "bodies and spirits," giving the daemons a composite existence as shadows produced by the movement of celestial bodies, as bodies moving on shadows, and as the knowledge obtained through measurement of the distance these shadows travel or extend between celestial bodies. The description of transit is also, for modern studies of literary genre, the narrative feature most clearly identifying this as a work of science fiction. The Daemon's detail in linking lunar transit to physical causes suggests, in this context, the mixture of science and the fantastic that marks science fiction, perhaps justifying Bozzetto's identification of the text as a missing link. But, the theological implications of this description of transit lead us away from the purely satirical intentions of Lucian and other precursors. At the same time, the mystical causes underlying scientific explanations, represented most specifically by the polysemy of the daemon as a motif that signifies the intermingling of spiritual and scientific concerns, shows us that Kepler uses science and fiction in a combination completely at odds with that employed by Verne and de Bergerac.

Most importantly, this section of the narrative demonstrates the personification of motion and suggests that motion, conceived by Kepler as an extension of divine power, serves as a dominant motif in the *Somnium* as it calls to mind a dense "network of physical and social ideas, natural and ideal concepts"[61] constitutive of allegory.

THE PHILOSOPHY OF MOTION

In contrast to Kepler, Galileo's conception of motion as an energy state leads him toward discussions of inertia derived in part from "a creative reordering of previously scattered physical and mathematical insights gained with difficulty by medieval scholars."[62] However, this concept also prevented him from viewing motion as a process, or considering the impact of one physical body on another. Thus, in his *Dialogue,* he only recognizes Kepler in order to ridicule Kepler's suggestion that the moon affects the tides.[63] Kepler's obsession with the moon struck Galileo as overly mystical, leading Galileo to discount Kepler, regardless of the accuracy of Kepler's conclusions. However, on the subject of the lunar influence of tides, Kepler notes that "The moon is a body akin to the earth," a conclusion that enables him, in his *Commentaries on the Motions of the Planet Mars,* to postulate the attractive power exerted by the moon on the water in the seas. In the *Somnium,* he also explains this process, but, in keeping with the conceit of the allegory, refers to the daemons as agents that facilitate the influence of the moon on the water. Thus, "if the daemons [. . .] undertake their work when the moon is favorable, its presence in the shadow will aid their efforts with the magnetic pull of a kindred body."[64] His explanation of this concept suggests both the gravity (which he defines as "a force of mutual attraction, similar to magnetic attraction"[65]) of the moon and an alternate, but equally significant, mystical force aligned with the moon. This explanation likewise affirms Neoplatonist theories stressing the existence of planetary souls while also pointing to the proliferation of meanings enabled by the allegorical form of the text—the form of Kepler's explanations allows him to suggest the complex and harmonic interplay of physical phenomena and spiritual forces at work in a single eclipse or wave.

The essentially mystical conception of celestial harmony formulated by Kepler, then, affirms harmony as the dominant force underlying the design of the cosmos. The secret of this harmony lies in the proportions of celestial bodies; harmony, then, remains a theological concept, so that,

> Képler affirme, en outre, que les proportions de tous les corps célestes ont été calculées en fonction d'un spectateur terrestre [. . . .] La conaissance que nous pouvons prendre de l'harmonie du monde depuis la terre correspond donc au sens 'intenté' de ce monde.
>
> Kepler affirms, further, that the proportions of all of the celestial bodies have been calculated dependent on a terrestrial observer. The knowledge that we can perhaps take of the harmony of the universe from the earth corresponds, therefore, to the "intended" meaning of this universe.[66]

Geometrical relationships, as well as movement, help to express the shape of this mystical guiding force. In the *Nova Astronomia,* this translates into his physics as a belief that motion must always be a process with a direct cause:

> a mathematical point, whether it be the center of the world or not, cannot move and attract a heavy object [. . . .] Let the physicists prove that such a force is to be associated with a point, one which is neither corporeal nor recognisable as anything but a pure reference.[67]

He describes physical points, points of reference, the point of the spectator observing the movements of the cosmos, and these positions are based on the correspondence between the spiritual and physical universe.

Borrowing quite directly from Neoplatonist conceptions of the universe, Kepler supports the belief in an *anima motrix* or moving soul.[68] Celestial bodies also exist, of course, as spiritual bodies. The earth is included in this formulation as well, prompting Kepler to affirm that "there exists in the earth also a soul."[69] The concept of the *anima mundi* or world soul animated Neoplatonism. But recall Kepler's interest as well in motion. The force caused by the movement of a celestial body can likewise testify to its spiritual vitality. Movement is spiritual, precipitated by Divine processes. The sphere that moves of its own volition is not merely an object moved by the power of God. Instead, it has been endowed with a vitalizing force that enables such movement.

There remains, then, a correspondence between the shape of the universe and the symbolic function of that shape so evident in the Neoplatonist model. The relationship between the earth and the sun, for instance, corresponds to Kepler's search for the symbolic function of celestial mechanics. As early as the *Mysterium Cosmographicum,* Kepler sought to articulate the concepts undergirding this relationship that "connect[ed] the periods of the planets with their distances from the sun."[70] The concepts that guided him on this search can be seen as new and revolutionary or as vestiges of older Neoplatonist influences. Caspar, for instance, describes Kepler's concept of movement as the result of his intuitive genius:

> No less significant is the idea which guided him in his search. It is the new thought that in the sun there is situated a force which produces the planet motions, and which is so much the weaker, the further removed the planet is from the source of the force.[71]

But the idea of a solar force producing planetary motion gives physical explanation to a Neoplatonist concept.

The mystical basis of motion and the significance of the sun as a symbol of Divinity influence Kepler's search. It is not so much a new thought that celestial bodies influence one another. Instead, Kepler's translation of a supernatural concept into the language of a celestial mechanics informed by Copernicus marks the significance of his discovery. Thus, even as Kepler articulates a theory that seems so clearly separate from medieval models of the cosmos, this theory merely reinterprets medieval philosophical concerns.

The concerns underlying the heliocentric theory are not so distinct from those of the geocentric universe. Kepler's conceptualization of the sun as center owes as much to empirical observation as to a belief in the sun as a powerful spiritual body. As Hallyn notes, "Le soleil n'est pas un site d'observation, mais un lieu de convergence par rapport auquel s'harmonisent les mouvements 'venant de toutes les provinces du monde'" [The sun is not an observation site, but a place of convergence for that which harmonizes the movements "coming from all the corners of the earth"].[72] The sun, a powerful but ultimately unattainable position from which to observe the universe, maintains, for Kepler, its connotative significance in the Neoplatonist universe as a place which engenders spiritual harmony, despite the process of desacralization inaugurated by the nova astronomia.[73]

The Copernican universe translates these spiritual qualities into physical characteristics:

> In the center of all rests the sun. For who would place this lamp of a very beautiful temple in another or better place than this where from it can illuminate everything at the same time? [. . .] And so the sun, as if resting on a kingly throne, governs the family of stars which wheel around. (Copernicus 16: 527–8)

Rather than abandoning or transforming the earlier formulation of the sun as a powerful spiritual entity, Copernicus in this passage personifies and ennobles the sun while suggesting the patriarchal power of the sun as it "governs" the "family of stars" it has presumably engendered. Paradoxically, as this passage suggests, the sun's new position allows it to see all while blinding those who return the gaze. The description of this unattainable and unassailable vantage point recalls *Paradiso* XXVIII, 16–21, summarized by Piero Boitani as follows: "He is looking into God's light. Dante had already glanced at it from far away: in the ninth heaven God had appeared to him as a mathematical point radiating such light that it blinds the eye."[74] The sun, ultimately unattainable, holds the ultimate place of observation. Indeed, Hallyn notes that "le soleil ne correspond pas au bon point de vue, mais joue le rôle d'un corps par rapport auquel les mouvements des planètes se recomposent *pour un spectateur situé sur la terre.*" [The sun does not correspond to a good point of view, but plays the role of a body that directs the movements of the planets for a spectator situated on the earth].[75]

Kepler's new universe, reliant as it is on concepts taken from the older Neoplatonist model, necessitates a different method for verifying its shape. On the most basic level, this casts doubt on the veracity of sensory experience. The heliocentric universe does not seem "right" when one considers sensory experience. This problem of the apparent untrustworthiness of sensory experience likewise attests to the preoccupations of Mannerist art. Indeed, Hallyn makes the point that this quality of the new astronomy echoes the problematic relation between artist and representation featured so prominently in Mannerist art. Panofsky's comment rings true for both astronomy and art: "that which in the past had seemed unquestionable was thoroughly problematical: the relationship of the mind to reality as perceived by the senses."[76] This is a key point for Hallyn's connection of Mannerism and the *nova astronomia* as disciplines that, while separate, reflect one another.

This point likewise applies to the connections between the concepts expressed in the *Somnium* and its attendant narrative form. The problems that Panofsky mentions in the previous statement can apply to both science and art. But they can also apply to the scientist producing art. It is not so much that scientists and artists, working in their separate disciplines, arrived at these problems of representation only in their major field of study. Instead, this disjunction—the relationship of the mind to reality as perceived by the senses—is also evident in Kepler's selection of the lunar voyage as a vehicle for his proof of Copernicanism. On the most rudimentary level, the text begins with the presupposition that the senses—of earth and moon dwellers alike—are insufficient for documenting the shape of the universe. Kepler's detailed explanations of the biology of the moon and the customs of its inhabitants may also be understood as indicative of this relationship.

But with a text like the *Somnium,* we see both science and narrative problematize the relationship between reality and the senses. The narrative form of the somnium, the *narratio fabulosa,* privileges this sensory dislocation. As a form with its own history in philosophical exploration, as I discuss in "The Process of Stellification," it likewise provides an ideal vehicle for the scientific concepts explored by the text. The combination of the dream and the journey removes the subject from the tyranny of conscious reality. In the dream, distanced from the immediacy of sensory perception, Kepler's narrator is able to discover a reality verifiable through the Daemon's speech: transcribed in Icelandic, filtered through a book, and appearing in a dream, which appears in a book, this information is distorted in its representation.

The theological concept of the earth as a *mundus propter nos conditus* [world constructed for us] is revealed as fiction. However, this does not mean that such a fiction is negative for Kepler. Instead, each world presents its inhabitants with a fiction that must be interpreted. Thus, it is customary to view Kepler's scientific work as "an uninterrupted reflection on the semiosis of a world whose figurative representation was undergoing a radical change."[77] But we must also consider Kepler's own choice of representative modes within the context of change.

The *Somnium* is not, then, completely divorced from its precursors in either form or content. We are not witnesses to the birth of a new genre. The new mode of expression is built from an interpretation of past sources, hence Kepler and Galileo's diverging viewpoints and writing styles. The selection of narrative form and the material portrayed by that form are connected. As Roger Bozzetto remarks,

> La naissance d'un genre différent au XVI siècle justifie, car des possibilités nouvelles de jeu avec de nouvelles réalités résultant d'un changement dans la conception du monde et des rapports entre le savoir et le monde se font jour: un nouveau type de savoir se propose, modifiant le champ conceptual—un nouveau paradigme.
>
> The birth of a different genre in the sixteenth century suggests, because of the new possibilities of play with the new realities resulting from the change in the conception of the world and of the connections between knowledge and the world of the time: a new type of knowledge proposes, in modifying the conceptual field, a new paradigm.[78]

He claims that the narrative possibilities of the game of fiction are closely tied to knowledge of the natural world. Bozzetto's conclusion that this new knowledge results in a new paradigm may seem a bit misleading, however. The new paradigm is merely a modification of the conceptual terrain: the perception of nature is still filtered through human consciousness. At the same time, Bozzetto's assertion that "À nouveau paradigme correspondrait donc nouveau type de narration" is of particular importance to our understanding of the conceptual framework informing this narrative. The new genre is inaugurated through its extension and reexamination of mythographic concepts and old allegorical forms.

The "scientific" revolution of the *nova astronomia* is clearly indebted to the mythographic concerns of the Neoplatonist universe. Thus, while Kepler describes the *Somnium* as a "plaything" or "hujusmodi ludicra," the motifs in this allegory "take textual form as extended tropes that rewrite the significance of their historical pretexts."[79] Even the philosophical implications of the circle demonstrate the extent to which, in Bruce Clarke's words, "literary allegorizing operates on the cusp of world-views" and "typically interced[es] between a problematic past and a present of cultural desire."[80] Nicolson's argument that the scientific voyage cannot be imagined without the supernatural voyage extends to the most elemental concerns of the supernatural voyager: the magician becomes the astronomer by breaking the magic circle.

Chapter Eight

Kepler's Allegories: The *Somnium* is not a Somnium

MYSTICAL CATEGORIES

Unlike earlier narratives of astral navigation, the *Somnium* employs imaginative narrative in the service of verifiable science; Kepler uses a narrative form associated with spiritual revelation to construct a thought experiment initiated from the dislocation of point of view necessary to create the selenography. The primary narrative device used to produce this dislocation of narrative consciousness is the journey. The narrative frame provided by the dream enables the journey in the *Somnium*. The dream, however, operates as a polysemic narrative space. Kepler borrows the journey motif, enabled by the narrative frame provided by the dream, to produce the necessary dislocation of narrative consciousness. The narrative space of the dream, while codified, as I explain in "The Process of Stellification" and "John of Salisbury on the Dream Book," likewise maintains an open space, an aporia, allowing the incursion of the unknown or uncategorizable. The many dream categories do more to point to the insufficiency of such cataloguing systems than to actually transfix and quantify the dusky clouds streaming from Somnus' cavern.

Still, such categories resonated for the medieval imagination. Disputes regarding the origin and attendant truth value of dreams stem from disputes regarding the barriers between the natural and supernatural, or between verifiable interpretation and ineffable vision. Thus, the status of the *Somnium* as dream evokes the question of the *Somnium* as allegory. An interrogation of medieval dream categories allows, then, for an exploration of this hermeneutically complex document. Kepler, in fact, continuously affirms the allegorical status of the *Somnium*. By allegory, he means, of course, traditional allegory, characterized by a limited series of binary correspondences. But, as every major contemporary critic of this work has indicated, Kepler's use of allegory is, instead, decidedly complex, and reveals the gradual shift to scientific language. Indeed, the polysemic qualities of Kepler's allegorical method impel its continued relevance for literary scholars.

But, is there anything that justifies contemporary attribution of the textual complexity of this narrative? Paul de Man's conception of allegory drives investigation of this textual feature. Indeed, de Man's theorization that "allegory becomes the property of language in general, a plot of figures redone and undone that is 'modernist' insofar as it echoes twentieth-century modernism's sense that it is cut off from the past and thus outside history" resonates strongly for the *Somnium* because it so prominently features "all figures [. . .] forged in time and subject to decay" as well as the process of this decay.[1] I argue that, to locate the origin of polysemy in the *Somnium*, we have to look no farther than the title. Allegorical polysemy, a relatively recent idea formulated by Franco Moretti, was, perhaps, conceptually unavailable to Kepler. Medieval allegorical theory posits Christian truth as an interpretive end-point encountered by all allegories. However, the multiplicity of Macrobean dream categories demonstrates that a fusion of categories was imaginable. By examining the *Somnium* in relation to these categories, I stipulate that the *Somnium* relies on the fiction of the dream as an essentially generic model. Likewise, this generic model, despite Kepler's maintenance that the *Somnium* can be easily deciphered in accordance with the interpretative practices of limited personification allegory, enables allegorical polysemy. Furthermore, the narratological analysis of Kepler allows us to consider, in greater detail, Ernst Cassirer's claim in *The Individual and the Cosmos in Renaissance Philosophy* that "even at those points where it seems to be freeing itself from the findings of Scholasticism, the philosophy of the early Renaissance remains bound to the general forms of Scholastic thought."[2]

I conclude this chapter by discussing the central object in Kepler's selenocentric universe: the Moon, a place that is never a no-place, that "is there for all to see in its various phases, and has always been there, has always been reached for, and has always been cried for [and that exists as] an almost tangible topos, particularly since the invention of an effective telescope."[3] Still, the transition from mythic to scientific modes of understanding the moon produces the interpretive complexity so readily apparent in this minor work. However, this complexity is, also, chiefly apparent to modern critics because of the position of the *Somnium* in a tale already told. Kepler does not consciously craft a new allegorical method in the same way that he does not set out to create "science fiction"; instead, this text exhibits multiplicity because of its position at the intersection of mythic and scientific consciousness.

In *Imagining the Unimaginable*, Lambert tends to view the dream space as a blank slate. The dream is a device which always serves to enable the imagination. In introducing the effect of the dream on the journey, she notes that "Kepler creates a sensory experience in which he invites his readers to participate in their imagination."[4] The first point to consider here is the intended audience of the *Somnium*. We cannot be certain that Kepler wished it to be widely circulated. Likewise, some would say that the text is incomplete. He continually added to the manuscript even to the end of his life. Still, Lambert is right to point out the interactive nature of the narrative. The compounded narrative levels produce a disorienting reading experience which can, as Lambert asserts, best be described as sensory and spatial.

The conceptualization of narrative space resonates for the *Somnium* because of the disorientation produced by each successive frame of this deeply embedded narrative. In discussing narrative space, Mieke Bal notes that

> Spaces function in a story in two ways. On the one hand, they are only a frame, a place of action. In this capacity a more or less detailed presentation will lead to a more or less concrete picture of that space. The space can also remain entirely in the background. In many cases, however, space is 'thematized'; it becomes an object of presentation itself for its own sake.[5]

This is particularly relevant to the conception of the dream as a landscape which determines the limits of the narrative. Though the dream as space implies openness and freedom, we see that Kepler constructs the *Somnium* in accordance with classical conceptions of the narrative possibilities open to the dream.

Lambert goes on to state that, because of the narrative element of the dream, "actual experience is superseded by experience in the imagination."[6] The dream as narrative device posits the existence of two separate worlds where narration occurs: within and outside of the dream. For Lambert, the act of dreaming, like writing, symbolizes the free play of the imagination. Imagination and dreaming are synonymous: "the dream [. . .] encompasses experience and representation as well as imagination."[7] Imagination, defined by Lambert as "experience and representation in the mind"[8] must remain distinct from experience and representation in the conscious realm. This is so because the act of imagining something results in a representation that is very different from that same thing outside of the dream.

For Duracotus, son of a witch and a fisherman he never knew, movement is a precondition of his lineage. Thus, as I have noted, Kepler thematizes space as movement. Duracotus' mother sells "little bags [made] out of goatskin, which she filled and carried to a nearby port to sell to the ship captains"[9] because of the weather-influencing magic of these bags. These bags recall the bag of "blustering winds" that Aiolus gives that perpetual wanderer, Odysseus. Although, unlike Aiolus' bag, Fiolxhilde's contains "herbs and linen cloth embroidered with various symbols."[10] The embroidered magical symbols that "power" the bag suggest, beyond our earlier exploration of this passage, the prohibition against writing imposed by Fiolxhilde. Thus, in Kepler's dream of a book, Duracotus, the author, tells the reader directly that "Her recent death freed me to write, as I had long wished to do. While she lived, she carefully kept me from writing."[11]

The dream, then, besides symbolizing the imagination, also provides a convenient means for distinguishing physical and mental representations. The narrative boundary of the dream maintains this distance between what are, for Lambert, two very different processes. But we must not forget that the dream in the *Somnium* is not an actual marker; there is no actual dream in the *Somnium*. Furthermore, both the imaginary dream and the realm of actual experience are depicted in the same narrative medium. This imaginary dream is part of the realm of the embodied imaginary: it is a text and "the experience of dreaming is at once real and fictitious."[12]

The classification of the dream as a text becomes problematic here for Lambert. She notes that Kepler treats dreaming and imagining similarly. Thus "dreaming could [. . .] be compared to a particularly vivid, unconsciously performed act of the imagination."[13] Here, the distinction between dream and text seem unclear in Lambert's argument. In the previous quotation, for example, we could define a dream as a particularly vivid and unconsciously performed imaginative act. Lambert appears to provide this definition in order to clarify the extent to which dreaming should be regarded as a subcategory of the imagination.

The dream is obviously a textual construct, a narrative device intended for a particular effect. As a narrative strategy, the dream calls attention to the context surrounding the dream. However, Lambert is not content to view the dream as merely an effect of the text. On this point, she diverges from Hallyn's assessment that the dream is, above all, "a written text, for which the appropriate operation is not reconstituting an experience but analyzing the meaning."[14] For Hallyn, the somnium is a mode consciously chosen for the presentation of the material concealed therein. The dream helps Kepler to provide a proof for his stipulation that the universe is not geocentric, but heliocentric. Its truth value as hypothesis overshadows its ornate and fantastic components. Indeed, in the period between the publication of Copernicus'

De Revolutionibus and Isaac Newton's *Principia Mathematica*, "the hypothesis had been reinscribed in a new discursive formation: that of 'science' as opposed to 'fiction,' the poetic."[15] Kepler, accordingly, reacts violently to the "absurd fiction" produced by Andreas Osiander in the preface to *De Revolutionibus* "that the phenomena of nature can be demonstrated by false causes."[16]

This quality informs the status of early modern scientific texts. At the same time, our assessment of the limits of the textual and imaginary cannot be determined merely by the generic forms utilized by early modern science writers. Lambert, thus, differs from Hallyn on her estimation of the boundaries separating textuality and lived experience. Hallyn's discussion posits that the *Somnium* necessitates an allegorical unpacking of the text using the methods of textual analysis familiar to his audience: those well-versed in the kinds of allegorical texts on which the *Somnium* is based. Lambert makes the claim that the *Somnium* is an "imitation of a dream" and thus "an imitation of an experience."[17] She is aware that the *Somnium* is a text; however, her thesis here posits a distinction between "felt and crafted experience"[18] that parallels the separation of the imagined and verifiable. To this extent, the *Somnium* details the genesis of the creative process. The imaginative elements of the text reveal the shape and direction of the scientific imagination. In this regard, Lambert builds on the main subject of Hallyn's book. She is interested in the *Somnium* as a document of the activity of the imagination. Her previous definition of dreaming as a particularly vivid, unconsciously performed act of the imagination must, then, be understood in this regard. The unconsciousness of dreaming is like the unconscious process of poetic and scientific inspiration. Further, the narrative construct of the dream allows for a conceptual means by which to distinguish inspiration, whether poetic or scientific, from the result of that inspiration.

Kepler takes pains to separate these through his documentation of "the coming-into-being of its dream narrative and the description of the lunar world as a creation of the imagination."[19] This point is central to Lambert's argument that the lunar dream presents a simulacrum of experience, as opposed to a purely imaginary hypothesis.

The lunar world, while clearly a product of Kepler's imagination, is not so distinct from his actual perception of the moon. He does, indeed, describe the moon as possessing not just mountains and other natural land formations, but actual cities, noting, in the "Geographical or, if you prefer, Selenographical Appendix" addressed to Paul Guldin, "Those lunar hollows, first noticed by Galileo, chiefly mark the moonspots [. . .] And in them the moon-dwellers usually measure out the areas of their towns."[20] This is not to say that he believed he was presenting a realistic depiction of the moon; he ends this Geographical Appendix by reminding Guldin that "these are playful remarks."[21]

Indeed, "there is a constant tension between the drift of the *Dream*'s rhetoric, which fashions the dream narrative as a description of an authentic experience, and the implications of its self-questioning."[22] However, his representation of the moon reflects his attitude toward his own lunar observations. Thus, the dream does not take place in a completely fantastic setting. The moon of the dream roughly parallels the actual moon to the extent that "Just as our geographers divide up the sphere of the earth into five zones on the basis of celestial phenomena, so Levania consists of two hemispheres. One of these, the Subvolva, always enjoys its Volva, which among them takes the place of our moon. The other one, the Privolva, is deprived forever of the sight of Volva."[23] More importantly, the dream-moon follows the same orbit as the actual moon and this orbit is described in accordance with the moon as it appears to earth dwellers. To this extent, Lambert is quite right to distinguish the *Somnium* as an attempt to simulate an actual experience. If we view the *Somnium* as completely textual, then we are less inclined to consider the results of Kepler's proof which rests at the core of the text to the world outside of the *Somnium*.

Still, Lambert does not deny that the *Somnium* is an artificial representation. The accretion of rhetorical effects in the narrative constantly recalls its own status as textual artifact. The distinction between these two ways of viewing the text gains its greatest importance when we consider the interpretive methods available as a result. As Hallyn argues, we must view the *Somnium* allegorically. He draws this conclusion from the title itself. A somnium is, of course, one of the categories of dream outlined by Macrobius in his *Commentary on the Dream of Scipio*.

Hallyn points out that we cannot simply view the title *Somnium* as indicative of any kind of dream. Instead, we must adopt the more technical language employed by classical and medieval oneiromancy. Kepler was clearly familiar with traditional works of dream interpretation, including Macrobius' *Commentary* as well as Cicero's *De divinatione* and Artemidorus' *Key to Dreams*.[24] Furthermore, Kepler's work providing astrological readings and dream interpretations for the Emperor ensures his familiarity with these works. These distinctions were still employed as late as the sixteenth and seventeenth centuries, and reveal the lasting influence of Neoplatonist speculation. Kepler's reliance on figures such as Macrobius and Cicero also betrays the scholastic reliance on *imitatio* as a discursive mode. The title, then, designates the content of the work as a particular kind of dream and evokes the text wherein the types of dreams are enumerated.

Kepler's *Somnium*, though, does not fit into the categories provided by Macrobius' *Commentary* as easily as one might think. Indeed, if we return to my discussion of dream categories in "The Process of Stellification," we will see that perhaps the *Somnium* is not a somnium at all. One might think that such a conclusion is irrelevant for our estimation of the importance of the *Somnium* within the history of early modern science. However, we must recall that early modern science was heavily indebted to areas of knowledge such as oneiromancy and astrology that we currently disregard.

These levels of classification join closely with discussions of the generic classification of this text. One of the key differences between Hallyn and Lambert's discussion of the *Somnium* hinges on its use of allegory. This textual marker—allegory—determines the truth values, the rhetorical elements of the text. Furthermore, viewing this text as a specific kind of allegory reveals the interpretive process necessary to "decode" its meaning. Still, one could contest that the *nova astronomia* merely inaugurated a continuous series of re-estimations of the central position of human consciousness, eventually necessitating the removal of the privileged Cartesian operator from the center of consciousness.

Kepler's own estimation of the truth of the dream hints at this gradual transformation of signification. Kepler, following Artemidorus, may have conceded that "we must judge as propitious everything that is in accord with nature, law, custom, occupation, names, and time."[25] But the *Somnium* also rewrites nature, revokes the law of authority. At the same time, the dream that is not in accord with nature may, according to Artemidorus, merely stem from "peculiar conditions which affect" the dreamer.[26]

Thus, the conditions surrounding Kepler the dreamer take on an added significance. Is his anxiety over the political situation in Bohemia a factor that could result in a dream that does not reveal the true shape of nature but is, instead, the expression of the disordered nature of Kepler's own mind? Artemidorus does not reveal the solution to this question: he has not written a prophetic book, but merely a handbook for aspiring prophets. He does, however, provide the enigmatic statement that "those who engage in rather secret activities dream of [. . .] a conspicuous chorus of stars at night as well as of the risings of the sun and moon,"[27] a pertinent assertion in that Kepler continuously reminds us of the occult connotations of his work. The discrepancy between Hallyn's estimation of the *Somnium* as textual and Lambert's that it is experiential is perhaps better understood then, through a discussion of the *Somnium* in relation to the dream categories formulated by Macrobius. These categories are, above all, concerned with the accuracy, the claim to truth, of the dream. The dream, any dream, is, for Macrobius, supernatural in origin.

The supernatural origin of dreams contributes to their truthfulness, and determines the extent to which they reveal hidden mysteries. This supernatural origin obscures an easy interpretation of the dream, however. In this part of Macrobius' schemata, every dream is a divine transmission.

There is still, moreover, the problem of physical stimuli. Most commentators also recognize the influence of the body on the truthfulness of dreams. Macrobius likewise affirms that the state of the body can impact the truthfulness of the dream. Aristotle, on the other hand, conceived of all dreams as the results of physical processes, thereby denying the existence of truly prophetic dreams. Macrobius' categories, and his views on the divine origin of dreams, are important because they were the most widely accepted.

Likewise, Kepler's title invokes one particular category taken from Macrobius' schema. The somnium is a true dream; unimpacted by physical contingencies, it ascends from the gate of horn to reveal divine messages to the dreamer. The differences between the gate of horn and ivory already speak to the genre of such a divine dream. The gate of horn is the origin of true dreams because horn, "when thinned, [. . .] becomes transparent."[28] Ivory, pale as the moon, "remains opaque."[29] The dream, when passing through the gates, perhaps becomes encased in these materials. It spins like a tiny planet toward the dreamer.

These materials, then, suggest interpretability, the ability to "see through" the shell and make out the form encased within this shell. The dream from the gate of horn is like the kind of allegory that we designate traditional allegory. The components, events, images of the dream are themselves transparent shells. Because they can be interpreted, the resulting interpretation is true. Dreams from the gate of ivory, however, remain opaque no matter how thin the shell. The true meaning of the dream is always hidden by its external frame. How would we define this type of dream in terms of literary mode? We cannot simply say that this dream has no meaning. Macrobius maintains that it is not meaningless, but instead provides a false or misleading meaning. Or, it features many meanings that are hidden within the opaque shell. All the dreamer can see is this shell, but the false dream still impels interpretation. Otherwise, Macrobius could not designate it as false.

A dream of, say, a rabbit with the wings of a bat is neither true nor false until we try to figure out what this image could mean, what dark errand our rabbit-bat has been sent to complete. The false dream is, then, like an allegorical text that suggests multiple interpretations. It does not set the value of one above the others. Or, it does not provide a meaning that can be resolved through recourse to typology or biblical allegoresis.

The somnium, of all the categories, is perhaps the most difficult to classify as true or false. It deliberately "conceals with strange shapes and veils with ambiguity the true meaning of the information being offered, and requires an interpretation for its understanding."[30] As I have pointed out in "The Process of Stellification," its definition is particularly ambiguous. Unlike the other categories of true dream, the somnium requires conscious activity. The dreamer does not know what the dream means until after the dreamer awakes. The content and meaning of the oraculum and visio are revealed in the dream space. Of the true categories of dream, the somnium is the only one that requires the conscious effort of someone who is not dreaming, exerting its power in both the dream space and waking reality.

At first glance, Kepler's *Somnium* seems to fit in quite well with this categorization. Of course the *Somnium* is a somnium. It also "conceals with strange shapes and veils with ambiguity": hence the narrative layers, the talking Daemon, the people of the moon. The footnotes attest to the necessity of interpretation. Outside of the narrative, they are also outside of the dream. For Kepler, these footnotes are integral to the understanding of the mysteries contained by the dream. The *Somnium*, because it features an authoritative figure who provides instruction (the Daemon), is also an oraculum. We could further classify it as a visio in that the image of the solar system in the dream is, indeed, "true."

As I have argued, though, such somnorific multiplicity is relatively unimportant: Macrobius likewise points out that the dream of Scipio was categorizable through all of the levels of true dream, despite the varying claims to truth of these true categories. But Macrobius' schema also sinks the *Somnium* into the murky domain of the insomnium, or false dream. The external frame of the *Somnium* supports such a reading: anxiety about the future causes the insomnium. The dreamer enters into an uneasy sleep through contemplation of waking life. The influence of waking reality factors heavily in the dream.

The *Somnium* begins with exactly this kind of situation, invoked in the very first sentence. The quarrel between the Emperor and the Archduke provided very real anxiety for Kepler. His position as astronomer was dependent on the whims of the Emperor. Distressed by this conflict, Kepler seeks answers in Bohemian history. He comes upon the story of the heroine Libussa, known for skill in magic. As the model for Fiolxhilde, she connects historical personages and dream characters. The dream is, as is established in the narrative, clearly influenced by Kepler's anxieties.

Above all, the *Somnium* is a text and not an actual dream. Kepler's anxieties and the historical context surrounding the production of this text add another level of interpretive possibility. As such, these categories remain useful because they coincide with the categories for defining and classifying allegorical texts.

The main distinction I have drawn between true and false dreams, and the supernatural origin of such dreams, leads us then to the question of textual classification. Still, the dream categories are useful in our consideration of the *Somnium* because it relies on the dream as a rhetorical strategy for producing an allegorical text. The dream as topos is a mechanism for producing allegory. Because dream and allegory are so closely intermingled, the categories of one inform our interpretation of the other.

However, the critics who have written on the *Somnium* disagree on how the *Somnium* operates as an allegory. I contend that this is because, in the same way that the *Somnium* expresses multiple dream categories, it also expresses different kinds of allegory. As an allegorical composite, it therefore suggests multiple interpretations. These types of allegory result, in part, from the position of the text between the supernatural and the scientific. This is why every critic dismisses the validity of Kepler's insistence that we read the *Somnium* using the strategies for reading traditional allegory. The critics suggest different ways of reading this text. As a polysemic allegory, then, it signifies in ways that distinguish it, narratologically, from the source texts on which it is based.

Still, Macrobius was not the only source to influence Kepler's ideas regarding the dream. A direct contrast with Macrobius can be found, for example, in Aristotle. The Aristotelian position, as I have noted, denies the divine origin of dreams, and views the dream as a series of random images. It is difficult to completely ascertain the influence of Aristotle's theories of dreaming on Kepler. On the one hand, Aristotle's theories of dreaming helped to change the late medieval attitude towards the divine origin of dreams.[31]

Furthermore, as in many other fields of knowledge in Kepler's time, "the authority of Aristotle, who, since the peak of scholasticism had reigned over both the philosophical and the physical domain, had in a manner been intensified."[32] We know that in astronomy, Kepler, like his contemporaries, sought to justify his findings through appeal to Aristotle's venerable authority. However, Kepler's strong Christian faith may have led him to deny the full extent of Aristotle's dream theories.

Aristotle's dream theories are most distinguishable from Macrobius' schema due to Aristotle's denial of divine dreams. While some writers (Kruger provides Adam of Buckfield's commentary on Aristotle's *De divinatione* as an example) followed Aristotle's lead, "a full acceptance of Aristotelian positions [. . .] was the exception rather than the rule."[33] The main reason for this was, of course, that "the denial of divine dreams clashed with biblical and patristic authorities."[34] As we have seen, the dream books were both popular and attributed to biblical figures. However, this does not imply that all Christians, let alone Kepler, affirmed the existence of divine dreams.

Kepler's attitude towards astrology hints at his complex viewpoint regarding such subjects. In response to the astrology of Helisaeus Roeslin, Kepler stresses the hidden component of supernatural influence: "that the heaven does something in people one sees clearly enough: but what it does specifically remains hidden."[35] So, while the heavens do exert an influence on behavior, they "do not compel, they do not do away with free will, they do not decide the particular fate of an individual, but they impress on the soul a special character."[36] In such discussions, Kepler is quite serious in his maintenance that a supernatural cause exists for both conscious and unconscious human behavior. Furthermore, he stresses the hidden nature of supernatural influence; the true draw of the heavens is veiled, one component among many. Still, despite the supernatural drive of Kepler's curiosity, Cassirer argues that, when interrogated, the complex early modern view toward astrology, as evidenced by another response to astrology, Pico della Mirandola's "Treatise Against Astrology," reveals the nascent components of the scientific method.[37]

At the same time, astrological prophecies provide him with a genre which he can parody as he performs his official duties. In his calendars, for example, "he prophecies because he is not averse to playing with the rules of astrology, but he immediately adds that one should not depend on prophecies."[38] These examples demonstrate a juxtaposition of genre and prophecy that informs our discussion of the *Somnium*. In the calendars or works such as his *Stella Nova*, he addresses an audience that responds best to astrological prophecies.[39] In such works, Kepler provides his audience with prophecy, and likewise enjoys the imaginative freedoms and writerly exuberances allowed him by this genre. At the same time, he does not want to go too far with this genre. Though he makes many prophecies and prepares astrological readings, he is well aware of their limitations.

His writerly dispute with Helisaeus Roeslin provides one example of this. Roeslin was an astrologer as well as physician-in-ordinary to the count palatine of Veldenz as well as the count of Hanau-Lichtenberg in Buchsweiler in Alsace.[40] The conflict between Kepler and Roeslin, taking the form of a dialogue in, among others, Roeslin's *Discurs von heutiger Zeit Beschaffenheit* and Kepler's *Antwort auff Röslini Discurs*, was based on Kepler's treatment of Roeslin's prophecies in his *Stella Nova*. Kepler critiques the exaggerated form of Roeslin's prophecies, and ridicules Roeslin's reliance on astrology.

However, he does not completely discount or dismiss astrology or prophecy themselves. Instead, as I mention above, he stresses the extent to which the stars conceal mysteries, or sing as a chorus for the initiate. The harmony of the universe, a subject so important to Kepler's work in astronomy, likewise exerts a considerable influence on astrology.

Kepler also had to defend astrology. We must not assume that Kepler was a luminary leading science away from astrology while the rest of the educated people of his time saw armageddon behind every harvest moon. At almost the same time as he was engaged in his dispute with Roeslin, he also composed a defense of astrology against the decidedly anti-astrological stance of Philip Feselius' *Discurs von der Astrologia iudiciaria.*[41] We might be inclined to think that Feselius based his argument entirely on an empirical critique of astrology. And, while he does fault astrology for its uncertainty, he also resorts to purely theological arguments denying the validity of astrology. As with dream interpretation, astrology was not caught between rational and mystical explanations. Instead, the arguments for and against these practices were heavily informed by a mixture of empirical observation, mysticism, and citations from biblical and classical authorities.

Kepler's reply to Philip Feselius is a defense of astrology with the unwieldy title *Tertius Interveniens, das ist Warnung an etliche Theologos, Medicos vnd Philosophos, sonderlich D. Philippum Feselium, dass sie bey billicher Verwerffung der Sternguckerischen Aberglauben nicht das Kindt mit dem Badt aussschütten vnd hiermit jhrer Profession vnwissendt zuwider handlen* [*Tertius Interveniens, that is warning to some theologians, medics and philosophers, especially D. Philip Feselius, that they in cheap condemnation of the star-gazer's superstition do not throw out the child with the bath and hereby unknowingly act contrary to their profession*].[42] We see Kepler then not completely discounting astrology. Indeed, by referring to theologians, medics, and philosophers, Kepler evokes those areas of knowledge privileged by the Seven Liberal Arts. These professionals who critique astrology hinder the significance of their own professions.

Further, Kepler's term for astrologers, "Sternguckerischen," includes astronomers. Depending on how we read his term, in fact, we can read Kepler's warning as conceived along disciplinary lines—this text becomes, then, the warning of a scientist with a background in the Trivium and Quadrivium who sees the limits of those disciplines and their adherents. Indeed, while Kepler began his career as a teacher of mathematics in Graz, later, under Brahe's tutelage, he embraced "the role of astronomer as active reformer and natural philosopher, supported by princely patrons and free from the pressures of university teaching."[43]

The "Sternguckerischen" is not quite a "star-gazer" who, enraptured by the welkin at dusk, idly absorbs the majesty of the night sky. Instead, the German verb "gucken" denotes "to peep," or "to peer." In fact, the noun "guckloch" refers to a peephole or spyhole, something that allows only a very focused and concentrated gaze. This emphasis on peering and peeping suggests the modus operandi of the astronomer—reliant on the telescope's peephole and the pin prick of the camera obscura—and not the work of the astrologer unable to distinguish astral bodies from mythic functions.

His dialogues on astrology take the form of a witty repartee with colleagues employed in similar positions. He uses the genre of astrological speculation while maintaining an ironic detachment from complete identification with astrology. At the same time, Kepler sees in such mystical exploration some of the key factors that motivate his scientific research. He cites the accuracy of the Copernican theory in both his defenses and dismissals of astrology. Furthermore, in his defense of astrology against Feselius' critique, Kepler compares the act of Divine Creation with the quest for knowledge of the human intellect. The tension between the supernatural and the scientific becomes resolved, then, in the quest for the *signaturis rerum*, with the extent to which the signature of the Divine can be read in the things of the world. Thus, for Kepler, "to inquire into nature is to trace geometrical relationships."[44] But we cannot stop there: for Kepler, to inquire into nature is, significantly, to trace textual relationships as well.

The book of the world, signed by the hand of the Divine, is then like the book of the word. Both conceal meaning; furthermore, the methods for reading the book of the world were clearly influenced by the methods for reading the word. Both necessitate an allegorical process of reading. Allegory, in this sense, necessitates the fusion of the supernatural and supranatural that we find so closely intermingled in the *Somnium*. Hallyn has classified the *Somnium* as an allegory. This classification, like the classification of the *Somnium* as a somnium informs not just our interpretation of the work, but the processes by which we are to interpret the work. The text certainly appears to take the form of an allegory. But what kind of allegory do we mean when we refer to the *Somnium*?

REWRITING THE MOON

The historical circumstances surrounding the text, most specifically hostile reaction to Copernicianism, seem to suggest the argument that it was necessary for Kepler to include his proof within the narrative shell of the cosmology. Kepler himself laments that "Everybody suffers his own injustice."[45]

He mentions the controversy surrounding *De revolutionibus*, wryly observing that its detractors "believe that this work must not be read unless the motion of the earth is first eliminated. This amounts to the same thing as saying that it must not be read before it has been burned up in flames."[46] However, "the choice of the dream genre seems to have been made for its positive possibilities and not as the act of prudence that one might imagine, in the defense of theses not acceptable to some authorities."[47]

Also, we must not make the mistake of ignoring the significance of this narrative shell. As Mary Baine Campbell warns in "Alternative Planet: Kepler's *Somnium* (1634) and the New World," "it is disrespectful of fiction, even fiction at its most skeletal, to think of it as a palliative add-on or pedagogical strategy."[48] Campbell points out that such speculation as is included in the *Somnium* was not necessarily prohibited. She stresses that "institutional denial of Copernicanism was not *total* and did not lead *invariably* to censorship."[49] Works housed in allegorical frameworks, on the other hand (Campbell cites Bishop Godwin's *Man in the Moon* (1638) and Cyrano de Bergerac's *Voyage dans la Lune* (1657)) seemed to encounter problems despite their fictional frame. As a counter example, Campbell provides *A discovery of a world in the moon* by the Anglican bishop John Wilkins. This work, "based on 13 propositions supported by logic, authorities, and some credible data" presented in prose did not result in controversy.[50]

The type of allegorical reading necessitated by the *Somnium* is a source of dispute among current scholars writing on this text. The type of reading suggested by Kepler himself, however, is almost completely discounted by every critic. Kepler suggests in his notes that we read the *Somnium* as an allegory reflecting Kepler's own life. Lambert summarizes some of the possibilities offered by Kepler:

> Duracotus could stand for Science, his mother for Ignorance or purely practical experience. The father, who died very old when Duracotus was still an infant, could signify Reason; he remains anonymous because Ignorance cannot know the identity of Reason. Moreover, the fact that Duracotus can only start to write after his mother's death shows that not until Ignorance has been dissipated can Science reveal to the public the deeply hidden causes of things.[51]

The characters are, given Kepler's interpretive possibilities, mere personifications of solidly recognizable, uncontested categories. Following Kepler's suggestions, then, we are faced with the *Somnium* as a limited personification allegory. Furthermore, Kepler's suggestion that we read the *Somnium* in this way is more of a testament to the influence of Neoplatonist sources. The model of the spheres was not the only important influence of Neoplatonist writers. Instead, this method of presentation, evident in, among others, Alain de Lille, Bernardus Silvestris, Dante, and Chaucer, also marks the *Somnium.*

None of the contemporary critics writing on the *Somnium*, however, seem receptive to this suggested reading practice. The *Somnium* as depicted by these critics, by contrast, offers more than limited personification allegory. Of these critics, Hallyn is the most sympathetic to Kepler's method, writing that this method constitutes a "completely classical" approach to oneiromantic interpretation. Lambert takes Hallyn to task, asserting that "this kind of allegorical interpretation is hardly permissible,"[52] and arguing that Kepler's own interpretive method must be ignored. Timothy Reiss likewise dismisses this interpretive strategy, suggesting that "the allegorical interpretation is explained in the *Somnium* as a merely convenient way to pass from the expression of self to the presentation of the scientific material."[53]

While limiting as a method, Kepler's analysis of his own work is in keeping with established approaches to the fabulous narrative. However, Hallyn's assessment of the *Somnium* as allegory is not as unqualified as Lambert makes out. Instead, "the allegorical sense does not function as origin and end of the fiction, as something on which the fiction depends entirely."[54] The levels of embedded diegesis reflect the various levels of available allegorical possibility. Such a reading contains the key for the examination of allegory as a changing mode that I propose. The "allegorical sense" of the text encompasses more than the individual narrative strands of the *Somnium.* The competing generic registers of the narrative are as imbued with an allegorical consciousness as are individual characters. Thus, the narrative strands of the *Somnium* allegorize the generic registers in which they are composed.

The fates of characters, if interpreted as personifications, escalate and finally address language. Such linguistic self-reflexivity announces itself in the barrier between knowing and not knowing. Fiolxhilde "never told [Duracotus his] father's name." Likewise, upon Duracotus' arrival at Hven, Brahe "began to ask me many questions. These I did not understand since I was unacquainted with the language except for a few words."[55] In the first example, Kepler notes that this detail has a simple allegorical explanation: "untutored experience or, to use medical terminology, empirical practice is the mother who gives birth to Science as her offspring."[56] Later in the text, he slightly alters the allegorical tag phrase attributed to the mother: "if you grant [. . .] that the mother of science is Ignorance, and that the father in reality is Reason, of course this father is quite properly either not known by that mother or concealed by her."[57] In the second example, Kepler comments on the linguistic barrier between Brahe and Kepler. This barrier, surmounted after only several weeks, sets the stage for Duracotus' acquisition of another language—that of scientific discourse.

In fact, Brahe's attitude toward Duracotus affirms the philosophical value of this new discourse. While Duracotus "did not understand" Brahe's barrage of questions, Kepler affirms Brahe's conversational method, noting that "It was the habit of that true philosopher never to stop asking questions, acquiring information, valuing such reports highly, thinking about them repeatedly, and applying them to the laws of nature."[58] Tycho, opposed to Ignorance as "that true philosopher," addresses the untutored Duracotus from two points of remove: language and mode. We are not to presume that Danish itself is the language of true philosophy, or Science, or that Duracotus' native Icelandic itself constitutes a mythic or pre-scientific mode of discourse, although the history of these two languages could accommodate such a reading. Instead, Kepler presents Duracotus' experience on Hven as one of linguistic immersion: his acquisition of "knowledge of the most divine science"[59] arises from his ability to process and reproduce languages.

Upon Duracotus' return to his mother, "she exclaimed that now she was ready to die, since she was leaving behind a son who would inherit her knowledge, the only thing she possessed."[60] If we maintain an analysis of this statement based on the limited series of binaries available to personification allegory, we encounter an impasse. How can Duracotus, a figure of science, returning from years of study and application, inherit the knowledge of Ignorance? How can his acquisition of the scientifically-infused Danish of Hven provide him with the linguistic inheritance of Icelandic (a language which is still notoriously resistant to the expression of technological and scientific concepts)? At this point, the method of traditional allegoresis leads us astray.

As readers eager to understand Kepler's meaning, we turn to the footnote intended to explain this ancestry of knowledge. We encounter, instead of one of Kepler's conversational explanations, a cryptic and unattributed poem:

> A charioteer dreams about a carriage,
> A judge about a legal fight.
> The wealth you seek by day,
> You acquire at night.[61]

The first two lines of this poem appear to refer to Lucretius' discussion of dreams in *De rerum natura*, to which I refer in "The Process of Stellification." The subject of dreams evokes the profession of the dreamer. Also, textual allegory reaffirms the essentially allegorical nature of human society: people are allegorical agents defined by careers and social positions.

Notes like this one "express a lack of certainty about which allegorical key to adopt."[62] Hallyn concludes that the confusion surrounding interpretive strategies is, itself, a strategy intended to highlight the juxtaposition of explanation and interpretation. Thus "Duracotus and Fiolxhilde can be *interpreted* as incarnations of universals, but also *explained* as products of the imagination, which engenders their names, their personalities, their actions, according to a particular rhetoric that is not mastered."[63] The boundaries of these discursive strategies are not so easily identifiable. Further, the presence of each category merely highlights Kepler's position, and the position of the *Somnium* in particular, within these two competing ways of reading. Hallyn emphasizes this disjunction because it strengthens our understanding of the role of the non-rationalized imagination in the thought process of a figure often placed within a linearly constituted history of the development of modern science. Hallyn's emphasis of the simultaneous dissociation and juxtaposition of two divergent types of reading that are "not aligned on an axis at the same textual depth but correspond to signifying configurations linked to the adoption of different points of view"[64] then helps us to see Kepler's allegory as a text able to be read in different ways through these divergent interpretive systems.

The personifications in the *Somnium*, then, help to suggest an alternative way of reading the text. In other words, the suggestions Kepler provides for how to read the personified figures in the *Somnium* do not offer us the entire field of possibility. However, neither should the suggestions Kepler provides be completely ignored. Thus, the method of reading the personifications in a generalized mythic manner helps to reveal an alternate set of readings that suggest the mythologizing of the scientific imagination. Hallyn provides an analysis of the figure of Duracotus to emphasize this juxtaposition of methods, or the extent to which one seemingly poetic method suggests a scientific interpretation. In the *Somnium*, then, along with the elements necessary for the proof of heliocentrism, we find as well "an entire network of narrative and descriptive elements [that] correspond [. . .] to the projection of mythical associations: mother, island, water, herbs are inscribed in the framework of traditional lunar themes."[65] These motifs, then, recall the moon as a distinctly mythographic icon. They exert an influence similar in scale to Kepler's classical sources. Through these motifs, we are to view the moon not only as a celestial body that, given its proximity to Earth, serves as a convenient setting for a completely rational demonstration of a legitimate theory of the shape of the solar system.

Instead, the presence of these motifs reminds us that the moon is also a source of a certain kind of imaginative power; its invocation bestows a wisdom beyond logic. The moon is, we are reminded, the source of lunacy; however, this lunacy, seen as an overturning of logic and order, provides us with the imaginative inversion necessary to demonstrate an even greater demonstration of the logic of the celestial order. Furthermore, these motifs assert the influence of the moon on the sublunary world. This is a scientifically valid concept which Kepler intuited through these motifs.

These two kinds of reading co-exist in the *Somnium*, then, and the markers of mythographic interpretation likewise point to Kepler's scientific rationale for compiling the *Somnium*. This juxtaposition of interpretations is also evident in the relation between Duracotus and Fiolxhilde as recounted in the text. Fiolxhilde, the herb woman, is clearly linked to these motifs which evoke the spiritual energies of the moon. Above all, these very real associations led to Frau Kepler's trial for witchcraft. Duracotus, as son, aspires to the knowledge of the moon that his mother so clearly embodies. In tracing this relationship, Hallyn notes that "Duracotus becomes and remains the symbol of what Bachelard calls the *will to intellectuality*."[66] He uses an example from Bachelard's *Psychoanalysis of Fire* that compares a child to Prometheus, and extends this comparison to Duracotus. Indeed, Hallyn notes that if we "substitute the young Duracotus for the child [in Bachelard's example] and the lunar figures of the mother and the sachet of stolen herbs for the solar figures of the father and fire, [...] we have a similar situation."[67] The relationship of father/sun and mother/moon is clearly archetypal. But, as Hallyn's notion of competing strategies of allegorical disambiguation suggests, in Kepler's *Somnium*, there is also a decidedly scientific component to this relationship, this genealogy of influence, with which we must familiarize ourselves.

As I have pointed out earlier, the sun for Kepler is not merely a figure symbolizing the characteristics attributed it by classical mythology. Hallyn's connection of Duracotus and Prometheus is telling as well; Prometheus is a bringer of wisdom. He not only makes the world of human culture possible, but also exists between the realms of gods and men, both as a bringer of fire and, after he is punished, chained to a mountain where he awaits the daily arrival of the eagle that will devour his liver. But his intermediary status is further complicated by the fact that, technically, he surpasses both gods and men. As a Titan, he is older than the gods. The wisdom he represents surpasses the limited interests of the Olympian gods. In this way, the Titans are more like forces of nature than the gods themselves, despite the association of certain gods with aspects of nature. While the gods control nature, the Titans embody the ultimately uncontrollable energy of the natural world.

Prometheus, or fore-thought, presages Duracotus who "represents continuity and progression, becoming."[68] Indeed, for Cassirer, the Prometheus myth "supplies the vital force in the battle against astrology and against the *Weltanschauung* of late antiquity, and finally decides the victory over them."[69] But Prometheus also has a brother, Epimetheus, or after-thought. Prometheus brings knowledge; his punishment even suggests the painful burden of knowledge in becoming. The penal code of classical mythology records a second punishment for the theft of fire. This example strengthens Hallyn's likening of Duracotus and Prometheus. While we are all familiar with Pandora's box of evil and ill will, it should be remembered that Zeus ordered Pandora created and married off to Epimetheus as punishment for Prometheus' theft. The contents of Pandora's box, and Pandora herself as a representation of the dark traits associated with femininity (particularly when these dark traits, impulsive curiosity among them, are basically mirror images of the light traits, like wisdom and the pursuit of truth, which we associate with Prometheus) bring us back to the solar/ lunar dichotomy with which I began this discussion. Duracotus is like both Prometheus and Epimetheus, at turns forward thinking and progressive, at others looking to seemingly outdated modes of knowledge associated with sources of evil magic.

Thus, returning to the last lines from note 32, we notice a reversal that does not support a positivistic view of the supremacy of a value- and rhetoric-free scientific discourse. "The wealth you seek by day" could refer to the astronomical knowledge pursued under Brahe's tutelage. Brahe, a luminary figure like Apollo, shines with the fiery radiance of numeric data. Fiolxhilde, by comparison, lurks in the preternatural shadow of witchcraft and sorcery. Despite Duracotus' active daylight search, however, he only acquires knowledge through shadow. In the most literal sense, of course, no astronomer can hope to study or understand the stars during the day; the act of stargazing necessitates the envelope of night. This logistic condition of celestial observation forces the astronomer into a new symbolic framework. The pursuit of knowledge, symbolized by light, the sun, the clarification provided by reason and fact, can only attain fulfillment in darkness, through communion with the moon, that orb of lunacy. Most significantly, this series of binaries (sun/moon, light/dark, wisdom/intuition), as pursued by Kepler, leads to a new set of characteristics that upset such clear and "natural" binarism. The sun, clarion of Neoplatonist philosophy, and an important emblem even for Kepler, illuminates truths that de-center its own previously unquestioned truth value.

In Fiolxhilde's supernatural understanding of the natural, Duracotus embraces a baroque fabulation that surpasses the mechanistic order of

Brahe's universe. Thus, for Duracotus to fully reconcile Reason (his father) and Empirical Practice (his mother), "he will have to be engendered a second time."[70] As a character in the *Somnium*, and second father to Duracotus, Tycho Brahe permits Duracotus to communicate with his mother; after his induction into true science, he is able to exist with his mother as "members of a spiritual community."[71] Yet this spiritual community manages to surpass the possibilities available to Brahe, with his immense household of astronomers. Brahe, who "allegedly [. . .] always wore full court regalia while observing because, most likely, that lent distinction to his pursuit,"[72] maintains a class-centered distance from the purely elemental communion achieved by Fiolxhilde who, as a result, claims access to a greater knowledge than that gained from Brahe by Duracotus.

Such a reversal clarifies that the *Somnium* "is not allegory in the proper sense."[73] Hallyn's presentation of the roles of the characters in the text indicates that we must extend our interpretation of the text beyond the narrow biographical level of meaning suggested by Kepler. Instead, we must view the allegory as itself engaged in a dialogue on the limits of science at the time of its composition. The revaluation of sun and moon, logic and intuition, mother and father, and any number of binaries attests to the continuation of the hand of the Divine on the written page. A reversal of natural correspondences described by Kepler through allegory, but ultimately derived from empirical observation, necessitates a transformation of the interpretive strategy necessary to decode that allegory. In this way, the *Somnium* reveals the crises of knowledge, the juxtaposed levels of mysticism and science, that characterize the development of early modern science.

The reluctance on the part of critics to designate the *Somnium* as just an allegory stems from a recognition of its modal hybridity. Of course, perhaps there is no such thing as a pure allegory, a text untouched by surrounding discourses and competing theories of knowledge. But with the *Somnium*, critics recognize a text that shows its author involved with these competing sets of significatory possibility. The juxtaposition of these systems does more than hint at the multiple ways of reading this text. Instead, in the *Somnium*, both sets of interpretive possibility are fully formed. Cosmological model and interpretive process inform one another. Recognizable hierarchies imposed by a Neoplatonist model impel a reading of motifs that differs from the same set of motifs regarded from a heliocentric, or even selenocentric, viewpoint. They certainly interpenetrate one another; even contemporary science is not as completely separated from culture, philosophy, and ethics as its most fervent technocratic proponents would lead us to believe. Kepler himself cannot be subjected to such a simple binarism as a scientist interested in a dubious mysticism or as a theologian (he studied for the clergy before switching to mathematics) with recourse to an excellent body of observational data. Instead, Kepler, like the *Somnium*, evokes at least two fully formed bodies of interpretive possibility. The allegorical method which played such a large role in Neoplatonist theories of knowledge and proto-science is used here in the service of a new discourse with its own range of interpretive possibility. Thus, in the *Somnium*, there are not two competing levels of allegory (the scientific and mythographic), but rather, to use J. Hillis Miller's phrase, two allegories, and at least two complete methods of reading these allegories.

TRIBES OF THE MOON

Paxson elaborates on the variety of interpretive processes we can bring to the *Somnium*. He uses Hallyn's reading to imply that the *Somnium* is "a self-reflexive advertisement of the isomorphism between mathematical systems with their geometric buttressing and literary systems such as the dream allegory, also buttressed by imaginary visual analogues."[74] The various narrative levels, charted by Hallyn as a series of "geometrically configured Chinese boxes"[75] emphasize the *Somnium* as a narrative representation of a geometric proof. By examining the *Somnium* with the tools of narratology, then, we note that this text, even in its form, is not an aberration from Kepler's other works. Instead, the preponderance of embedded narratives, each encapsulated within the other, suggests more than anything the *Mysterium Cosmographicum* which seeks to demonstrate the harmony of the solar system by inscribing the celestial bodies within geometric forms. Despite the fact that the *Mysterium Cosmographicum* "still gets scorned as an enthusiastic and misdirected opus redolent of Platonic or Pythagorean mysticism,"[76] its well-known reliance on embedded celestial bodies informs the *Somnium*.

The diagrammaticality of the *Somnium* suggests as well the competing allegories at work in the text. Paxson notes that scientists disparage the *Mysterium Cosmographicum* for its Neoplatonic mysticism. However, he demonstrates that a later work such as the *Nova Stereometria* also relies on the thema of embedding. And this despite the position of the *Nova Stereometria* as a "mature mathematical opus endorsed by 'modern' science" because it offers an important theory for the calculation of areas and volumes under curved surfaces."[77] Temporally, and in its claims to scientific authenticity, the *Somnium* exists between these two works. The Neoplatonism of the *Mysterium Cosmographicum* is also expressed in the *Somnium*, even as, on other levels, the *Somnium* operates as an allegory of becoming-science.

The interpretation of characters in the *Somnium* that Hallyn offers becomes, then, extended by Paxson into an estimation of this disjunction between Neoplatonist mysticism and scientific rigor. The thema of embedding contributes to the status of the *Somnium* as an "allegory of the phenomenology of space and time."[78] Thus, for Paxson, the *Somnium* "presage[s] through the customary language of allegory the semiotics of mathematical cognition."[79] Furthermore, the dream allegory is the form best suited for this goal, as it necessitates divided narrative levels and, thereby, distinct interpretive registers. Paxson uses Brian Rotman's conceptualization of the process of mathematical cognition to point out the extent to which the rhetorical strategy of narrative embedding parallels the geometric embedding that, while based in Neoplatonic mysticism, also enables significant developments for modern science. Rotman employs the topos of the dreamer in order to describe the process of mathematical cognition, effectively dividing the mathematician into two levels of consciousness, which he likens to sleeper and dreamer. The "dreamer awake" is the mathematician as a person; the "dreamer asleep" is the agent who navigates the symbolic realm of mathematics. Paxson suggests that Rotman's formulation turns out to be "frankly uncanny" in its applicability to Kepler's *Somnium*. Partly this stems from Rotman's reference to the mathematician as dreamer for "more than any other narrative topos, the theme of dream texture evokes the language of medieval and early modern allegory, the didactic, literary mode most steeped in phantasmagoric, hallucinatory, surreal, or oneiric experience."[80]

The dream provides a psychological space removed from waking reality and ideal for contemplation. It is not unusual that correspondence should exist between the process of mathematical exploration and the genre of the medieval dream allegory. Indeed, Hallyn notes that "dream, book, conversation with a supernatural being: these three modes of presentation correspond to the first three types of what Andrè Festugière, in his study of Hermes Trismegistus, calls the 'literary fictions' of the logos of revelation."[81] The dream as a motif is closely connected to revelation. The mystical connotation of the dream journey does not diminish with the development of modern science, however. Instead, as we note in the *Somnium*, the act of mystical revelation enables *logos*. Thus Duracotus can only converse with Fiolxhilde after his tenure under Tycho Brahe. Brahe, an exemplar of *logos* in the purely Aristotelian sense, provides the skills which actually reconcile Duracotus with Fiolxhilde, who perhaps can be equated with *pathos*, a purely intuitive and remarkably overpowering conception of the forces shaping the cosmos. Duracotus can only interact with the Daemon after his reconciliation with Fiolxhilde: the moon of the Daemon does not merely reflect the Neoplatonist sun, however. Instead, the moon reflects and distorts the natural order of solar wisdom.

The *Somnium* marks more than a transition in science history. Instead, the text, as allegory, expresses the slow decay of time (a concept inherent to allegory, as best expressed by Paul de Man's "The Rhetoric of Temporality"). Kepler's *Somnium* marks the birth of a new variety of cosmological journey, but it also signals the end of the narratives which presage the lunar voyage. The scientific proof underlying the text leads to a generic reassignation of texts like Plutarch's *De facie* and Cicero's *Dream of Scipio* which contribute to the narrative frame. By this, I mean that, while the *Somnium* inaugurates science fiction, it also fictionalizes its sources; Plutarch and Cicero's texts can no longer be read as valid possibilities. The truth claims of twelfth century cosmologists regarding the harmony of universe and man (as expressed, for example, in the *Cosmographia*) is ultimately discounted.

Lambert recognizes this result of the *Somnium* in her discussion of its allegorical qualities. The account of the history of allegory with which she ends her chapter speaks as well to the condition of Kepler's allegory between methods of interpretation. She recounts the story of Plutarch's *De facie*, highlighting its similarities to the *Somnium*. Of course, Kepler himself was well aware of these correspondences, remarking in his notes to the *Somnium*, "every time I re-read this book by Plutarch, I am exceedingly amazed and keep wondering by what chance it happened that our dreams and fables coincided so closely."[82] In the *De facie*, one element in particular, the presence of Cronus, illuminates the significance of its influence.

Banished by Zeus, Cronus sleeps confined in a deep cave, served by oracular spirits or daimones. Every thirty years, the inhabitants of a nearby island send a shipload of people who will worship or engage in the study of philosophy and astronomy. Cronus, a powerful figure in the *De facie*, provides the inhabitants of the island with prophetic revelations. While the daimones also serve as oracles, "the prophecies that are greatest and of the greatest matters [the spirits] come down and report as dreams of Cronus, for all that Zeus premeditates Cronus sees in his dreams."[83] Cronus' position, both in this narrative and in Greek mythology in general, is telling for Kepler's lunar allegory. Like Prometheus, Cronus precedes the Olympians. The youngest of the Titans, Cronus was also their leader. His defeat of Uranus marks a mythographical boundary. This moment rends the abstract timelessness of the Greek creation. Prior to this act, there is chaos; the emergence of Gaea from chaos; Gaea's creation of Uranus, the mountains, and the sea; Gaea and Uranus' parenting of the Titans, Cyclopes, and Hecatoncheires; and Uranus' attempt to hide his children in Gaea's body, effectively preventing their birth by stuffing them back into the womb. The juxtaposition of this event with Cronus' slaying of Uranus suggests the boundary between timelessness and temporality. Like Cronus after him, Uranus sensed death in beginnings: living children engender dead parents. Cronus' method for slaying Uranus hints at this paradoxical pairing of birth and death; he castrates Uranus with a huge scythe. The severed genitals, bloated and rotting on the salty sea, produce a foam from which Aphrodite emerges.

This castration results in the symbolic separation of the earth and the heavens. Cronus begins time by cleaving space, establishing it as something out there, removed from but giving shape to earth and its inhabitants. Cronus is, significantly, a god connected to dreams; his dreams are the truest because they presage Zeus' own thoughts. Evocative of time, Cronus, or *chronos*, "sees in his dreams all that Zeus premeditates."[84] Lambert uses the image of Cronos in the *De Facie* to comment on the relevance of the *Somnium* in a historical continuum. She ends her article with the statement that "the work of astronomers may bear fruit in time so that Kepler's lunar allegory can be exchanged in time for a literal account."[85] This statement, however, suggests a clear break between the allegorical and the literal. The form of the lunar allegory loses its relevance once the allegory has been properly interpreted.

This point is very important to the theory of allegorical motifs and interpretation that I am developing in this study. In traditional allegory, the events of the narrative serve as a shell within which meaning is concealed. The meaning illuminates these events and, at the same time, replaces them. Lambert suggests a variation on this process that is closely tied in to a historical progression of knowledge. This progression, suggesting as it does a shelf-life for interpretations, runs counter to the timelessness of the mythic references she uses to assert her point. The rhetoric of temporality, instead, suggests a continuous decay. The literal conceals, but can not entirely displace, the trace of the allegorical.

The story of Cronus, as myth, exists beyond time. This is the case even though the myth is concerned with the temporality of mythic prophecy. The symbols that dominate this myth parallel those used by Kepler. Both narratives, as Lambert notes, feature:

> different levels of narration, the multiplicity of textual references, the island, the years spent serving a god, the astronomy, the daemons, the opposition between the stranger's experience of a physical journey and the dream visions or imaginary journeys arising from Cronus' confinement, the combination of sitting in the dark and thus not seeing (Cronus in his cave) and 'seeing' in a dream, finally, the dream itself, the dream about the Moon that is.[86]

On the one hand, these similarities point to Kepler's reliance on the *De facie* as source material. On the other hand, we are faced with two narratives which, while built from similar motifs, necessitate entirely different methods of interpretation. This is the case because of the last element in Lambert's list of similarities: "the Moon that is." Interpretation transforms the moon that is or, by extension, the universe that is, as objects. Readers do not merely arrive at new interpretations of narrative motifs. Instead, the motif—in this case, the Moon as re-presented by Kepler's lunar allegory--changes the field of interpretations made available by the narrative model in which it appears.

One could argue that traditional allegory itself is not as unidirectional as postulated by, for example, proponents of the symbol over allegory as a preferred means of signification. In "Defining Irrealism: Scientific Develpmment and Allegorical Possibiity," I discuss the complexities inherent in the interpretation of traditional allegorical motifs. Such interpretation was not simplistic because the same object could assume contradictory meanings. For example, the lion can symbolize either absolute good or absolute evil; the dualism calls up the *in bono*/ *in malo* cast of allegorical signs according to St. Augustine or Bede. However, this symbol, like others within the Christian allegorical framework, was limited to theological interpretation. Umberto Eco, in *The Limits of Interpretation*, summarizes that:

> in order to understand the meaning of the facts told by the Bible, Augustine had to understand the meaning of the things the Bible mentions. This is the reason for which medieval civilization, extrapolating from the Hellenistic Phisiologus or Pliny's *Naturalis historia*, elaborated its own encyclopedic repertories, bestiaries, herbaries, lapidaries, *imagines mundi*, in order to assign a symbolic meaning to every piece of the furniture of the "real world." [. . .] The work of medieval commentators was to provide rules for a correct textual disambiguation.[87]

However, despite the multiple possibilities available to a single motif, these possibilities still operate within a single field of possibilities. The lion, for instance, can be construed as either good or evil; however, both interpretations stem from a single Christian world view. The interpretations of this motif do not suggest an alternate philosophical framework. Even the four levels of allegoresis employed in traditional allegory operate in relation to one another. There is no philosophical or ideological break between the literal, allegorical, moral, or anagogical interpretations of a single motif. For Dante, this is a result of the polysemous nature of allegorical figuration.[88]

Instead, they operate like the separate layers of a palimpsest, amplifying or supporting one another.

The situation of allegorical interpretation differs for the re-estimation of the moon provided by Kepler's *Somnium*. We could identify the moon through each interpretive level offered by traditional allegoresis. But recall that Lambert clearly distinguishes the Moon as symbol from the moon as an actual celestial body because the symbolic moon, particularly as employed in the *Somnium*, calls on a wide body of mythographical and philosophical source material. This results in the contradictory character of the text, so that demonology and magic set the stage for a defense of Copernicanism. Hallyn summarizes this oxymoronic intermingling of source material as follows: "The *Dream* contains this paradox: while one of its proclaimed goals is to promote the new concept of a heliocentric universe, the fiction that overlays this defense borrows some of its materials from ancient geometric cosmology."[89] Ancient geometric cosmology provides the symbolic range available to the moon. But the range of these associations does not extend to all possible readings of the moon. Instead, these associations contribute to the possibility of interpretation removed from these same associations. During this period of transition, then, the actual moon exists simultaneously in Neoplatonist and heliocentric universes; it wavers, or oscillates between both, like a biblical sign now *in bono*, now *in malo*. The moon of the heliocentric universe, however, is no longer guided by the fundamentally philosophical concepts that impel movement of the geocentric spheres. The traditional mythographic associations may still be used to refer to the moon, but, as Kepler shows, this moon belongs in a completely different universe.

Of course, a fundamental problem mars the validity of my previous statements. If the moon of the heliocentric universe can no longer be described through myth, then why is the selenocentric moon so resolutely populated with denizens of a mythographic imagination? Hallyn suggests that Kepler maintained the wide body of mythic associations in the text because "the geocentric universe, scientifically unacceptable as it had become, nonetheless remained more satisfactory in terms of imaginative associations, identifications, and correspondences."[90] For Hallyn, this does not really present a problem for how we read the text. He does suggest the importance of this overlap, noting that, "In Kepler's *Dream*, two totally different forms of apprehending the world briefly coexist, one presented in terms of the other."[91] We can imagine the orbits of two separate bodies, the Neoplatonist and heliocentric moons, briefly overlapping. The selenographic moon, visible from a stormy Bohemian vantage point, signifies the shadow of that eclipse.

These moons, composed of language, suggest academic disciplines that use different sets of terms to describe similar phenomena. For example, a narratologist and an information science specialist may employ different terms to refer to a single model used for organizing a body of information. However, the languages used for describing phenomena can also change perception of the fundamental qualities of a single phenomenon. Hallyn suggests this by evoking two worlds which "briefly co-exist." But this is not to say that the world of mythic language has disappeared. We can still find authors—Calvino, Borges, Kafka, for example—who rely on bodies of mythic knowledge that continue to resonate. Furthermore, the language we use in literary studies to describe these two worlds impels their continued and continuous coexistence. Hallyn does not use the past tense to refer to the period of overlap for the mythic and scientific moon. Instead, they briefly coexist in a continuous present itself indebted to the language of mythic representation.

Like Rosen, Hallyn docs not see any problem with the *Somnium* as a "sustained oxymoron."[92] Lambert, however, sees this response as "hardly satisfactory."[93] I do not quite agree with Lambert's conclusion that Kepler's scientific representation is not technically allegory but is instead an imitation of an actual experience. This supposes too much that the language of science solves for the disparity between reality and our representations of reality. But, as contemporary theorists and historians of science are quite clear to articulate, the language of science is as reliant on seemingly poetic devices—metaphor, analogy, and, of course, allegory—as the most unscientific of mythic narratives.

However, Lambert's dissatisfaction with Rosen's explanation of the synthesis of the mythic and scientific does allow for a more detailed understanding of the moon as motif and its impact on the lunar allegory as genre. The *Somnium* does not create a simulation of the lunar voyage. It also never escapes the resounding force of its mythic influences. While the text asserts the absurdity of geocentrism, it does so through a mystical selenocentrism that, despite "scientific" logic and reasoning, reveals the moon as an important spiritual body for the astronomer. "The Moon that is" lies embedded within a narrative so evocative of the moon that was, and remains, a textual creation. The *motif* of the moon, however, resignifies. The brief overlap of selenographic significations also suggests historically constituted limits and expansions of narrative representation. While mythic language shapes the moon of Plutarch, the shadow of the moon produced by overlapping languages can only be seen through a telescope equal parts polished glass, frogs' breath, and blood of the martyr.

Chapter Nine

The Speech of Daemons

THE MOON AS DREAM SPACE

In the "First Day" of Galileo's *Dialogue on the Two Chief World Systems*, Galileo provides an intriguing metaphor for describing the limits of the imagination. He offers the hypothetical example of a person who lives in a large forest filled with animals and birds but no aquatic creatures. Galileo suggests that this person would be unable to imagine anything about a lake or stream given his knowledge of an apparently body-of-water-less forest: "Even with the liveliest imagination, such a man could never picture to himself fishes, the ocean, ships, fleets, and armadas."[1] On initial observation, Galileo's example evokes the new worlds being discovered by explorers venturing into North and South America. The reports of these explorers, as William Shea reminds us in "Looking At the Moon as Another Earth: Terrestrial Analogies and Seventeenth-Century Telescopes," "are full of comparisons with what was already familiar to them either from personal experience or from reading and conversation."[2] The parallel between the physical exploration of the New World and the scientific exploration of astronomers like Galileo and Kepler comes as no surprise. These complementary revolutions expand the western concept of world, of what is real and imaginable.

Further, such comparisons of astronomy and exploration were common. While contemporary criticism tends to separate astronomy and travel, early modern disciplinary boundaries were not so fixed. On occasion, Kepler blamed these similarities for impeding sales of his books. In the 1621 edition of the *Mysterium Cosmographicum*, for example, "Kepler first complains that 'Cosmography' is sometimes used for 'Geography' and his work has on occasion been incorrectly catalogued by booksellers."[3]

A dialogue between the English astronomer Thomas Harriot and his friend Sir William Lower regarding Galileo's *Sidereal Nuncius* and its effect on their own lunar observations also reflects this rhetorical fusion of exploration and observation. After Lower connects his own observations to the surface features of the moon described by Galileo, Harriot writes to Lower:

> Me thinkes my diligent Galileus hath done more in threefold discoverie than Magellane in openinge the streights to the South Sea or the dutchmen that were eaten by the beares in Nova Zembla. I am sure with more ease and safetie to him selfe & more pleasure to mee. I am so affected with newes as I wish sommer were past that I mighte observe the phenomenes also. In the moone I had formerlie observed a strange spottedness al over, but had no conceite that anie parte thereof mighte be shadowes.[4]

This exploration really stemmed from the improved level of magnification made possible by the telescope. Still, Harriot compares observation to action. Galileo opens the path to the moon in the same way that Magellan "opened" the straits to the South Sea.

Or so it seems. Actually, neither Galileo nor Magellan *opened* anything; Galileo applied mathematics to the improved magnifications made possible by a new device while Magellan sailed around the tip of an inhabited continent. However, while many have gazed at the moon, none before Magellan had piloted a ship of European adventurers through these straits.

The shadows of the moon, necessitating recognition that the moon has depth and shape as well as a distinct geography, are not themselves self-evident to anyone with the proper mathematical training. Thomas Harriot, for example, was a mathematician, cartographer, and astronomer.[5] He had the ability to understand the mathematics underlying new theories to describe the appearance of the moon, having "solved the problem of reconciling sun and pole star observations for determining latitude, introduced the idea of using solar amplitude to determine magnetic variation, improved methods and devices for observation of solar and stellar altitudes, and derived a full numerical solution for the Mercator system of map projection."[6] However, he was perhaps more persuaded by Galileo's evocation of the moon as a planet, his rhetorical method, than by mathematical proofs. As Shea argues, the analogy of the moon as planet enabled a new way of looking at the moon.

In the letters between Thomas Harriot and Sir William Lower, Lower describes his problems with seeing the moon as Galileo suggests it be seen:

> According as you wished I have observed the moone in all his changes . . . [Near] the brimme of the gibbous part towards the upper corner

> appeare luminous parts like starres, much brighter than the rest, and the whole brimme along lookes like unto the description of coasts, in the dutch bookes of voyages. In the full she appeares like a tarte that my cooke made me the last weeke. Here a vaine of bright stuff, and there of darke, and so confusedlie al over.[7]

Despite his observations, Lower continued to "grop[e] for an apt description of what he sees."[8] Until Harriot explains Galileo's findings through the analogy of the moon as planet, Lower's "words fail him, and his imagination is tossed from the description of a coastline read in a Dutch travelbook to the memory of last week's pie. The terrestrial features of the Moon seem to cry out to be recognised, but Lower's vision is both overwhelmed and blurred."[9]

This difficulty with seeing the moon reveals the rhetorical significance of the writing of the new astronomy. While astronomers like Galileo and Kepler are frequently viewed as discovering new things, perhaps it is more useful to think of them as arguing in new ways. Galileo's command of geometry contributed to the forcefulness of his argument in *Sidereus Nuncius*, and his use of analogy and metaphor resulted in its persuasiveness. In this sense, Galileo relied on two tools: the telescope and the Aristotelian telescope, that equally illuminating and not entirely unrelated device created by Padre Emanuele in Umberto Eco's *The Island of the Day Before*. People did not come to believe that the surface of the moon was cratered and uneven only because of geometrical formulas.[10] Instead, they were persuaded once they had been taught by Galileo how to interpret their observations (made through the use of a scientific instrument that began its life as a toy). Shea's article, in turn, provides a clear historical overview of a particular problem. He traces Galileo's use of the telescope to the importance of Galileo's analogical argumentation.

But the quote from Galileo with which I begin this chapter of the person unable to imagine denizens of the deep or the deepness of the abyss remains problematic: Galileo himself describes things that he cannot really "see." The same is true for Kepler. However, both were able to build persuasive arguments from what they could see. This would be akin to a situation where Galileo's hypothetical forest dweller fashions a model of a fish out of branches and twigs. Or perhaps he hollows a tree trunk, fills it with water, and places a mirror at each end of the vessel so that he can imagine water stretching toward the horizon. In this sense, the forest dweller uses the materials around him to envision, and thereby describe, an Other world.

Indeed, like Galileo, Kepler explores the similarity between making and knowing through the allegorical conceit provided by the *Somnium*. The figure of the Daemon, simultaneously evocative of the scientific and the supernatural, and lurking at the center of the *Somnium*, serves as an example of the degree to which objective knowledge is determined by the form of its presentation. Likewise, the Daemon's ability to describe, dependent on the ability of an audience to process the ramifications of such a description, forces us to read the Daemon, a body constituted by speech, as a polysemic allegorical construction. The competing interpretations and valuations of this character as offered by Kepler hint at the centrality of the Daemon not just structurally, but also conceptually. The Daemon, then, serves as a site and embodiment of the competing discourses informing Kepler's attempt to reenvision the shape and order of the cosmos.

As the example of Harriot demonstrates, the verification of the Copernican theory relied as much on rhetorical presentation as on the geometric proof itself. At the same time, as I discuss in Chapter 7, Galileo and Kepler did not present ideas, or even conceptualize them, in similar manners. As a complex allegorical sign, the Daemon of the *Somnium* embodies and replaces the possibilities made available by the late medieval mystic tradition in ways unavailable to Galileo's *Dialogue*.

In thinking about the different modes of presentation involved in justifications of Copernicanism, I kept returning to the question of agency, of who describes or offers a description or explanation. We can not merely say that the narrative allows for the geometric proof offered in the *Somnium*. For example, James J. Paxson's concern with the narrative embedding so clearly evident in the *Somnium* suggests that the juxtaposition of embedded narrative layers coincides with the content (the fiction of the dream, the verifiable content of the footnotes, and the Daemon's speech) of those layers. As a former student of Paxson's, and admirer of his book on personification, I think that this argument doesn't go far enough. I wondered if we shouldn't instead think of the Daemon as a personified manifestation of this meeting of narrative levels. For Bruce Clarke, working from Angus Fletcher's theorizations of allegory, this confluence impels all allegorizing: "Mediating discontinuous eras and disparate realms, interweaving the threads of a textual web that would net the world in its mesh, the daemonic fictions of allegory weld a composite cosmos together."[11] The content of the Daemon's speech and the mode of its presentation form the body of the Daemon itself. Like Lucifer in the Ciudecca of the *Inferno*, the Daemon lies imprisoned at the center of the narrative. Still, in Kepler's narrative, this central position imbues the Daemon, like Kepler and Galileo themselves, with the role of teacher describing the observations possible from that vantage point. By this, I mean that Kepler's conception of the Daemon as a character plays into the role of the Daemon as a teacher.

Perhaps, in discussing imprisoned daemons, it is fitting to evoke Paul De Man who, in "'Conclusions': Walter Benjamin's 'The Task of the Translator,'" concedes that he "is not at all certain that language is in any sense human."[12] This comment suggests the monstrosity inherent in all language—in the gap between human as natural body and human as philosophizing and rational mind:

> de Man's rhetoric is marked by tropes conveying the specifically teratological threat that language's Other poses for the understanding. The 'monstrous' here functions as a philosopheme, a conceptual-figural strand linking quite disparate texts in unexpected ways and revealing a hidden coherence with regard to the relationship between language and the human.[13]

For this project, De Man's meta-theory of language serves to qualify the Daemon's speech as constitutive of a body that comes into being by revealing elements of "hidden coherence" between human experience and linguistic representation.

GEOGRAPHIC SPEECH

After the Daemon makes his didactic speech describing how humans are transported from the Earth to Levania, he goes on to speak of the form of Levania's provinces, commencing "like the geographers."[14] At this point, the Daemon indicates a shift in his mode of discourse. In order to discuss Levania, another planet, an alternate reality with moon people and Daemons speaking in blunt, hollow voices, the Daemon names a mode that will inform his discourse. This modal shift is of key importance to understanding Kepler's own creation and extension of the travel narrative and cosmological narrative in the *Somnium*. At the same time, such a modal shift also reflects our understanding of the Daemon itself as a character closely connected to this mode which it introduces. While the speech of the Daemon is associated with this narrative mode, the geographic tilt of the Daemon's speech is reflected elsewhere in the *Somnium*. Thus, as Mary Baine Campbell notes, "Kepler reproduces the new cosmographical context everywhere in both the narrative of the *Somnium* and its voluminous Notes. The latter are full of specific allusion to voyage literature, data from which are properly cited as if it belonged to the same technical literature to which Kepler's text putatively belongs."[15]

Campbell's observation here helps us to locate the *Somnium* in a generic context. At the same time, to return to my earlier point regarding personification and genre in the *Somnium*, there is considerable overlap between seemingly separate components of the narrative. Thus, as Campbell observes (and I have discussed in Chapter 8), the dreamed moon as setting is presented as a new place. This place is, as I argue, characteristically different from the mythic moon which preceded and influenced its creation. Kepler's description of Levania as an island most concretely expresses this transformative desire. To this extent, "The 'island' is the land form that functioned as a kind of mastertrope of New World topography, and that characterised the focus of classic voyage literature, especially where it spoke most directly to private desire: Columbus finds islands, and Thomas More, and Andre Thevet."[16] The island, however, as a mastertrope, is itself an allegorization of some set of qualities, and is essentially rhetorical in nature. The moon is depicted as an island, moreover, within a speech act, the speech of the Daemon. Through the footnotes, Kepler the scientist supports the Daemon's assertion of moon as island.

The speech of the Daemon, then, serves as an elocutionary act intended to provide a reexamination of an object, the moon as island. The speech, embedded within the *Somnium*, and itself encrusted with footnotes, must be considered in relation to a variety of narrative levels, including the notes and the surrounding narrative frame. The rationale behind the composition of the notes attests to their significance: "The Notes augment the scientific data already foregrounded in the main body of the narrative and defend various ludic moments against their ludicrous misreading in the events of his mother's imprisonment and trial."[17] At the same time, the Daemon who presents this speech exists as a character positioned between numerous interpretive possibilities, which we will turn to in this chapter.

Campbell sees the lunar voyage as a means of presenting "alternative worlds that offer most saliently the radical fact of Alternativity itself."[18] This concept of Alternativity, of Other-seeing and Other-being, expresses the potential available through the unique point of view provided by the lunar voyage: the moon is the island *par excellence*. Like the object of Roberto della Griva's constant obsession in Umberto Eco's *The Island of the Day Before*, the moon remains eternally unattainable. At the same time, the moon can be viewed in its entirety as a discrete object quite separate from all things earthly.[19] Yet, in a text that features as many narrative levels and narrators as the *Somnium*, the moon necessitates presentation and representation from varying perspectives.

These various perspectives allow Kepler to extend the relevance of the travel narrative and medieval cosmological text while reconfiguring the significatory value of these modes. The cosmological narratives of the twelfth-century school proposed a theory of proportion between the shape of the cosmos and the shape of the human soul. For these narratives, however, both ends of the spectrum are intangible properties. Neither the cosmos nor the human soul could be sufficiently or entirely mapped or imagined. Kepler relies on the narrative models provided by writers like Macrobius in order to do something very different. Here, the shape of the moon, and the certainty of the movement of the moon and the planets of the solar system, enables the possibility of a mapping of man as a creature in a material universe. Thus, the Daemon that speaks of "the earth to us humans" does so equally through the content of his speech and the various modes he employs in order to deliver his message.

DAEMON, DAIEIN, DAIOMAI

Kepler's Daemon is a complex allegorical personage. This character becomes a representative for a completely different conception of nature. Such a monstrous character,

> is born only at [a] metaphoric crossroads, as an embodiment of a certain cultural moment—of a time, a feeling, and a place [. . .] A construct and a projection, the monster exists only to be read: the *monstrum* is etymologically 'that which reveals,' 'that which warns,' a glyph that seeks a hierophant. Like a letter on the page, the monster signifies something other than itself: it is always a displacement, always inhabits the gap between the time of upheaval that created it and the moment into which it is received, to be born again.[20]

At the same time, the Daemon plays a key role in determining how readers value the generic modes that Kepler employs. The "revelation" of this *monstrum* involves not just the scientific conception of cosmology, but also the allegorical shift precipitated by such scientific discovery.

To begin, Kepler is quite aware of the allegorical connotations of the word daemon. In note 34, he writes that the spirits "are the sciences in which the causes of phenomena are disclosed."[21] He extends this link between science and magic as he introduces the speech of the Daemon. The Daemon synthesizes science and magic and also conceals this synthesis through the allegorical conceit. Thus, Kepler notes that the allegorical substitution of spirits for sciences "was suggested to me by the Greek word *Daemon*, which is derived from *daiein*, meaning 'to know' as though it were *daemon*."[22]

Kepler's problematic discussion of etymology, however, has been noted and contested by Rosen, who indicates that, "in deriving *Daemon* ("minor divinity") from *daemon* ("expert"), Kepler followed the etymological speculation of Plato."[23] Rosen refers specifically to Plato's *Cratylus*, and contrasts this with the etymology accepted by modern philologists, "who connect *daimon* ('the divinity') with the verb *daiomai* ('divide')."[24] In this etymology, the daemon is not a representative of pure knowledge. Instead, "a daimon was so called because he allotted their destinies to mortal men, cutting out their future for them."[25]

Kepler's choice of character, his mouthpiece for Copernicanism, did not simplify the reception of the message. As Campbell points out, Kepler's mother's trial for witchcraft stemmed from the description of Fiolxhilde's consultations with supernatural forces. Hostile political situations necessitate allegory. In discussing the history of the mode, Joel Fineman notes that "allegory seems regularly to surface in critical or polemical atmospheres, when for political or metaphysical reasons there is something that cannot be said."[26] Kepler attributes his difficulties to the ignorance of an audience whom he never intended: "I mean that these words fell upon minds which were dark within and suspected everything of being dark."[27] He scornfully rejects those who have attributed diabolical devices to his text. We can, as Rosen does, exclude Kepler's Daemon, then, from any kind of supernaturally malevolent implications. Thus, Rosen emphasizes that the term "was understood by many of Kepler's uninformed contemporaries to mean an evil spirit."[28]

Here, Rosen intentionally distinguishes between these uninformed contemporaries and Kepler, exemplar of a science informed by reason. However, it is ultimately impossible, and also unproductive to try to effectively seclude Kepler from the contemporary culture in which he was immersed. Instead, we can think of the different levels of society in which Kepler moved as a series of texts themselves involved in a complex process of synthesis. This is particularly significant in terms of my discussion of the Daemon, as I am interested in the Daemon as a textual body.

In *Les Technologies de l'intelligence*, P. Lévy offers a definition of hypertext which accounts for this cultural and textual synthesis of meanings. Thus, for Lévy, the hypertext is an "ensemble des messages et des représentations circulant dans une société" [unity of messages and representations circulating in a society]. He considers these as part of

> un grand hypertexte mouvant, labyrinthique, aux cent formats, aux mille voies [. . .] Le sont justement ces associations indues, ces métamorphoses, ces torsions opérées par des machines locales, singulières, subjectives, connectées sur un extérieur, qui réinjectent du mouvement, de la vie, dans le grand hypertexte social: dans la 'culture.'

> a grand, moving, labyrinthine hypertext, with a hundred formats, a thousand passages [. . .] It is exactly these unseasonable associations, these metamorphoses, these torsions operated by the local, singular, subjective machines, connected to an exterior, that reinjects the movement of life, into the grand social hypertext: into 'culture'.[29]

Rosen notes that, regardless of the extent to which he problematizes Kepler's etymology, Kepler's Daemon is more classical than diabolical. However, an examination of the *Somnium* reveals that this distinction is not self-evident. The "grand hypertexte mouvant" of cultural attitudes toward the Daemon cannot be supplanted from the theological, philosophical, and scientific impulses guiding Kepler's hand. Further, the definition of "daemon" that we accept impacts our understanding of the Daemon as a creature.

The Daemon, as mouthpiece and gate keeper between earth and the moon, is both a text and a body. Still, this body is also obscured from sight. It speaks only when it cannot be seen. In fact, we do not know anything about its physical appearance, or even its gender. Lambert makes the following observation regarding the gender of the Daemon:

> It may be worth pointing out that, in his translation of this passage from the Latin, Edward Rosen refers to the Daemon as female, thus setting 'her' parallel to Fiolxhilda rather than Duracotus. The Latin version leaves the question of the Daemon's gender open and thus offers a further source of ambiguity. The Greek word 'daimon', after all, can take both the masculine and the feminine gender.[30]

I use the pronoun "it" to avoid any confusion in this matter. The magical covenant accompanying the daemonic summoning necessitates that the Daemon not be seen. Duracotus and his mother cover their heads as part of the magical rite and are rewarded by the "rasping of an indistinct and unclear voice."[31] They cover their own bodies to access the Daemon in a manner recalling Huguccio of Pisa's derivation of the word *monstrum* from *mastruca*, a word referring to hairy garments or skins. Huguccio's warning that "'Who ever dresses himself in such garments is transformed into a monstrous being'"[32] strengthens the identification between Daemon, Duracotus, and, ultimately, Kepler.

Despite the hidden body of the Daemon, there remains a significant correspondence between the body and the voice. Kepler's description of the grating character of the Daemon's elocutions is accompanied by a footnote suggesting that "it is not impossible, I believe, with various instruments to reproduce individual vowels and consonants in imitation of human speech."[33] Kepler here refers to a talking machine well before its invention. However, the sounds produced by such a machine "will resemble rumbling and screeching more than the living voice."[34] Rosen questions Kepler's evocation of a talking machine and determines that "Undoubtedly [Kepler's] purpose was to suggest that what sounded like the rasping voice of a spirit might have been only an imperfect mechanical reproduction of human speech."[35] The conclusion Rosen offers, in conjunction with Barthes' view of the connection between body and voice as expressed in "The Grain of the Voice," suggests the kind of body attributable to Kepler's Daemon.

For Rosen, the reason for this attribution of daemonic rasping to an imperfect machine is perfectly clear: "In that case there was no spirit or daemon."[36] The absence of a spirit or daemon in the narrative would clearly alter the genre of the narrative: Kepler would no longer be the author of an occult fantasy featuring the fearsome personages of folk traditions. Instead, he would be the author of a science fiction narrative perhaps more modern than even Koestler could have anticipated. Thus, Duracotus' mother (and, for Kepler's more critical contemporary readers, Kepler's own mother by extension) could not be seen as a daemon-summoning witch. Instead, she would be heralded as an inventor of the highest order.

At the same time, such a ruse allows Kepler to further distinguish himself from Duracotus, who supposes the machine is the voice of a demon. In this sense, Kepler, like the Wizard of Oz, is aware of the machinery rumbling beyond the curtain. Thus, Duracotus falls prey to one of the "built-in traps for the superstitions and gullible," and supposes that "demons are talking" when, in fact, "art is copying magical tricks."[37]

The concept of the speaking machine is not, however, pursued elsewhere in the text. Even in the footnote where Kepler proposes the daemon-as-machine, he remarks that the sepulchral voice of the Daemon reminds him "pleasantly" of his colleague, Matthias Seiffart. Seiffart had an instrumental role in calculating the ephemerides of the moon for the year 1603. As a student of Brahe's and colleague of Kepler's, Seiffart was involved in many of the important conversations that resulted in the verification of the heliocentric theory. Still, besides having the voice of a daemon, "he was also affected by depression and mental illness, in which there was no place for relaxation."[38] Thus, by the end of the note, this voice is again firmly likened to a being and not a machine. Moreover, Seiffart, on account of his voice and depression, is not unlike a demon, or someone, at the very least, plagued by demons.

Kepler's notes on this subject present the voice as both physical and mechanical. As a physical phenomenon, the voice delimits the body of the

Daemon, qualifying a reader's imagination of the voice as the unrestrained bellowing of a creature unbounded by the laws of nature. To this extent, the qualities of this unnatural voice describe the unnatural body that houses the voice. At the same time, the voice may be purely mechanical, an artificial construct that sounds mysterious, but may be produced by a clever toy. If this is the case, the "daemon" could simply be a scientist playing a trick on Duracotus and his mother.

Either way, the voice calls attention to the unseen body that produces this voice, a relationship Roland Barthes calls the grain of the voice, and defines as "the encounter between a language and a voice."[39] The Daemon's essentially textual but unviewable body embodies "the very precise space (genre) of the encounter between a language and a voice."[40] As I have been arguing throughout this study, the encounter between a language and a cultural body finally able to speak that language results in a reconception of the genres that modulate and categorize languages, modes, and registers. To this extent, the body of the Daemon *is* the genre of the *Somnium*. The paradoxes of its voice and its body represent the competing discourses influencing Kepler's own estimation of the language the Daemon speaks.

Of course, complete languages do not hang in the air like dense clouds of miasmatic swamp gas, infecting the unwary traveler with linguistically-manifested psychoses. Expressions of culturally-guided embodiment, languages evolve in response to environmentally-supplied stimuli. However, languages also affect bodies in a manner similar to that employed by the fungal parasites that infect and control the nervous system of the ant species *Megaloponera foetens*[41]: Barthes' description of the voice as body, evoking as it does "the cavities, the muscles, the membranes, the cartilages [. . .] as though a single skin lined the inner flesh of the performer and the music he sings,"[42] suggests that the grain of the voice signifies "the materiality of the body speaking its mother tongue."[43] From this standpoint, that which can speak the Daemon's language inhabits the Daemon's body.

Kepler continuously likens the magical summoning of the Daemon to his own practices of astronomical observation. Still, Kepler's note on this voice suggests that even the Daemon is not to be viewed as an unimpeachably supernatural manifestation. The scientific experiment, like the magical ritual, is only effective if it is repeated, unchanged, time after time. Thus, like the ceremony performed by Duracotus' mother, "the corresponding feature in the teaching of astronomy is that the method is not in the least voluble or spontaneous."[44] Kepler does not draw this parallel between magic and science without providing an example that further clarifies his selection of the form of the magical ritual that appears here in the *Somnium*. Instead, he provides an example of the standard procedure he used with visitors when he worked in Prague:

> Whenever men or women came together to watch me, first, while they were engaged in conversation, I used to hide myself from them in a nearby corner of the house, which had been chosen for this demonstration. I cut out the daylight, constructed a tiny window out of a very small opening and hung a white sheet on the wall.[45]

Prefiguring Newton's *Optics*, Kepler describes a demonstration of the camera obscura (perhaps, for our purposes, best referred to as a camera oscura, or shadow chamber), its history also linked to witchcraft.[46] Through the citation accompanying notes 44, 46, and 47, Kepler ironically likens his preparations to those of a diabolical necromancer. Regarding his preparations, he asserts that "these were my ceremonies, these my rites."[47] In his description of these procedures, he also emphasizes his instrumental role in producing the demonstration. Furthermore, he notes that he hides from his guests, and that he, like the Daemon, produces knowledge, demonstrates mastery of earthly elements, while, at the same time, remaining invisible, remaining legible only through the medium of the demonstration itself.

The final part of Kepler's description of his magical rites further emphasizes the textual quality of this magico-scientific demonstration. The camera obscura produces a dark enclosure with an aperture through which light enters to form an image of outside objects on the opposite surface. Light filters through the opening and produces an image on the inside of the dark enclosure. In further describing his demonstration, Kepler asks the reader "Do you want characters too?" By this, he refers to the magical symbols accompanying the sorcerer's ceremonies. Because the camera obscura produces a reflection of the image, he was forced to write his message backwards so that it would be legible to his audience. Still, drawing on the daemonic lore of black masses, based on a rhetorical *inversio* identical to the inversion of earth and moon animating this text, Kepler comments "behold the magical rite" as he mentions that "the shape of the letters was backwards [. . .] as Hebrew is written."[48] Kepler's invocation of Hebrew is deliberate. He also uses the name "Levania" for the moon because "Moon" in Hebrew is "Lebana or Levana."[49] While he notes that he could also have used the word "Selenitis," he chose the Hebrew because "Hebrew words, being less familiar to our ears, inspire greater awe and are recommended in the occult arts."[50]

Furthermore, Kepler's evocation of magical symbols appearing mysteriously on the wall recalls another text concerned with dream interpretation and the occult arts: the Book of Daniel, which, as we have seen, is also linked to the genre of the allegorical dream narrative.[51] Kepler, however, does not merely liken these reversed messages to the right-to-left movement of the Hebrew alphabet; medieval European culture frequently linked the Jews to black magic, cannibalism, and other inversions and perversions of the Christian faith.

The structure of the camera obscura also emphasizes the connection between the magical and the scientific. The spectators are enclosed in a room which functions as the camera. While the image is projected into the room, those inside the room may have difficulty seeing outside through the small opening. Kepler positions himself, like the Daemon, on the outside of the camera. This is not to call into question whether or not Kepler was actually in the room as he demonstrated the visual illusions of the camera obscura to his houseguests. Instead, I mean to point out that, in this description, Kepler emphasizes his position as the creator of the illusion. While he was most likely in the box, his intellectual mastery of physical forms assumes a mystical presence outside of the room temporarily transformed into a camera obscura. Indeed, the mechanism of the camera obscura works in such a way as to maximize the supernatural atmosphere it invokes. Thus, "If a breeze disturbed the board outside, the letters inside wiggled to and fro on the wall in an irregular motion."[52] A scientific tool like the camera obscura suggested to Kepler the theological questions underlying the science of astronomy. The mysterious sensations produced by a scientific tool hint, further, at the promise of understanding the supernatural through empirical means.

DAEMONIC DYNAMEIS

In *Allegory: The Dynamics of an Ancient and Medieval Technique*, Jon Whitman adds a further dimension to our conception of the supernatural qualities of the daemon. He notes, most significantly for our consideration of the *Somnium*, the complexity of the relationship between psychic causes and cosmic forces as articulated in texts from late antiquity. The *Somnium*, building from traditional mythographic sources, also echoes this disparity. Whitman notes that "already in Plato and Xenocrates, there are intriguing associations between the opposing forces of the soul and the various daemones of the world, who mediate between the realms of spirit and matter."[53] Prior to this, the individual soul was not so closely identified with the processes of the universe. In philosophical texts of the first centuries AD, "a correlation increasingly develops between the elements of the soul and the 'powers' of the spiritual world as a whole."[54]

This disjunction corresponds with the increasing allegorization of the various features of the human personality. From a purely narrative standpoint, this correlation between the elements of the soul and the denizens of the spiritual world is most evident in terms of character and setting. The personification of these elements changes the possibilities of philosophical speculation. The *Psychomachia* of Prudentius, for instance, emphasizes the dramatic possibilities of texts speculating on the reconciliation of the individual soul and Divinity. The early Christian desire to "bring the soul to heaven"[55] achieves a literal solution through the personification allegory. Following the interpretive process necessitated by allegory, such narrativization "places the drama of the soul simultaneously on two planes."[56]

Whitman's analysis of this component of allegory is significant for this discussion because of the conclusions he draws about this 'turn inward' that characterizes early Christian allegory. For Whitman, "this turn inward has its inverse compositional counterpart, as the articulation of the soul expands outward into the world at large."[57] The turn inward, which results in a narrativization of internal mental processes, necessitates the creation of a landscape on which these psychological battles and journeys are enacted. This imaginary landscape, as it represents more and more the imagined workings of the soul, overlaps more and more with that most abstract, but still "real," landscape, the cosmos. Thus, "by the time of the *Cosmographia*, this process will become panoramic in scope."[58] The two-part structure of the *Cosmographia* explicitly emphasizes the correspondence between microcosmos and macrocosmos that appeared early on in Plato and Xenocrates. At the same time, this correspondence, expressed as philosophical concept in Plato and Xenocrates, becomes a generic model of textual organization in the *Cosmographia*, so that these concepts also become ways of dividing the text.

Whitman also draws the intriguing connection between the daemons of the world and the dynameis, or governing powers, of the world. Thus, "the dualistic tendencies of such discussions intensify in Plutarch, who at times speaks both of two guardian daemons for the soul and of two governing powers (*dynameis*) in the world."[59] These quotes point toward an identification of daemons with dynameis. Thus, according to Whitman, the term daemon becomes closely associated with the concept of natural or spiritual powers. This differs markedly from the previous definitions of *daemon* offered by Kepler (and Rosen). The daemon as power is neither the daemon as expert (Kepler) nor the daemon as divider (Rosen). Whitman traces this development of the meaning of the word in relation to the development of Judeo-Christian conceptions of the supernatural. Thus, "Perhaps the most telling development of this period is the tendency among writers from various traditions—Jewish, Hermetic, and Christian—to conflate different senses of the term 'powers' itself (*dynameis*): psychic forces, angelic influences, and divine attributes."[60] By this, Whitman means to point out that the two separate realms where daemons may exert their power—the individual consciousness, the natural world—are themselves conflated. He notes Origen as the writer in whose works these diverse possibilities emerge as a single phenomenon. Thus, Origen "divides both the soul and the world (and individual nations) between good and evil daemons—mediating powers (the evil ones associated with particular beasts and vices) who inhabit the air and help or hinder the soul's ascent to heaven."[61]

Here as well, the various psychic landscapes converge, guarded as they are by the same plethora of good and evil daemons.[62] The dominant allegorical topos highlights the struggles of the soul as it attempts to ascend to heaven. On the one hand, the daemon-plagued landscape the soul encounters is merely an attempt to represent a spiritual journey that is essentially unrepresentable, stemming as it does from the meditative withdrawal of the pious pilgrim from the external world. At the same time, the conflation of daemons with dynameis renders this metaphorical process more 'real.' Thus, while the temptations of hostile daemons may exert themselves entirely within the psychic sphere of the individual soul, the powers of such daemones, like the powers of their holy counterparts, are not limited to the individual conscience. These powers are, then, operative within both the individual and the universe. For Whitman, "Such conflations tend to incorporate the middle realm of the daemons within the framework of man or God. The 'powers' are in a sense elements of each, but at some remove from both."[63] The powers of the daemons can, then, work within the individual and the universe. However, the daemons exist, at the same time, at a remove from the individual consciousness.

Kepler's Daemon is certainly emblematic of this conception of the daemon. Kepler's literary sources attest to this. But in the *Somnium*, the idea of the daemon as a force or power also has repercussions for the idea of the natural and for the newly developing language for describing the natural that we would identify as the voice of modern science. The distinction between the daemon as expert or the daemon as power is of central importance here. The expert always exists at some distance from the subject of expertise. The ability to describe how something functions necessitates objectification: if something is described, it is described from some exterior vantage point. The voice of the Daemon alone certainly justifies such a conception of the range of the daemon.[64] The switch in mode from Duracotus' narration to the Daemon's learned discourse merits our conception of the daemon as an expert articulating his views.

APPARITION IN THE NARRATIVE FRAME

But is this final discursive mode—the language of the geographers—truly encapsulated by the narrative? I contend that its use, on the contrary, opens up the many narrative levels of the *Somnium*. To this extent, the innermost layer of the *Somnium* then points beyond itself, effectively leading to a new language that discards the encapsulating levels of the narrative as a snake sheds its skin. From the standpoint of traditional allegory, "the knowledge of things, as well as language, is essential to proper interpretation."[65] For Christian allegorists, as Augustine emphasizes in *On Christian Doctrine*, "the thing a text signifies should in turn signify another thing, until all signs eventually disappear in God."[66] In this way, Kepler reconciles a mathematically verified model of the cosmos that contradicts the Neoplatonist model through recourse to the discursive modes employed by Kepler's predecessors. Thus, for Kepler the theologian, a science based on mathematics can, like "the allegorical interpretation of scripture [,] expand beyond the text into the world at large, diverging radically from the initial correspondence between text and meaning."[67] The Book of Nature, a metaphor "itself derive[d] from the Latin Middle Ages,"[68] animates Kepler's discussion in a slightly different manner than current understanding of this metaphor would suggest. Instead, the Daemon's textual body parallel's Alain de Lille's formulation that

> Omnis mundi creatura
> Quasi liber et pictura
> Nobis est et speculum.[69]

This body, a microcosmic jigsaw composite of ideas and modes, demonstrates a correspondence with the macrocosm of the new universe; this correspondence is conceptually similar to the Neoplatonist theory of correspondence, but results in very different bodily forms. The deforming metadiegetic frames of the *Somnium* suggest the textual representation of this correspondence. James J. Paxson explores the subject of narrative frames in detail in "Revisiting the Deconstruction of Narratology: Master Tropes of Narrative Embedding and Symmetry." He notes that:

> As a rule, or in accord with what might be identified as an implicit narrative code, the innermost framed discourse or endodiegesis of the

> *Somnium* seems to grow ever more abstract and dense. In this final narrative voice, the voice of the professional geographers, the *Somnium* speaks the language of contemporary science, one given to empirical observations, much measurement, and geometrical or trigonometrical rendering.[70]

Of course, this discursive mode, the language of contemporary science, was, in Kepler's time, not itself encapsulated or codified. Indeed, this language was very much in a process of becoming. Science, as a discursive mode, was not entirely freed from the spiritual dimensions of previous attempts at macrocosmological knowledge.

The metaphysical description of the motion of the sun is further based on rhetorical ingenuity. The *Somnium* begins, conceptually, from a fundamental dislocation of point of view: "For Levania seems to its inhabitants to remain just as motionless among the moving stars as does our earth to us humans."[71] A moon-based perspective replaces the earth-based perspective which has yielded the geocentric concept. While the logic remains the same, the moon-based perspective transcends the boundaries of possibility, belonging neither to the geocentric, heliocentric, or Tychonian universes.

But the dislocation of point of view from the earth to the moon is not without precedent in Kepler's scientific work. After Tycho Brahe's death in 1601, Kepler was appointed Imperial Mathematician. In Prague, he began to work on the problem of planetary orbits using Brahe's calculations. However, as he complained in a letter, "'I would already have concluded my researches about world harmony, had not Tycho's astronomy so shackled me that I nearly went out of my mind.'"[72] In order to determine whether his calculations of planet orbits were correct, Kepler employed a device familiar to readers of the *Somnium*. He was confronted with the problem of verifying the various positions of the earth's orbit. Making these calculations from his vantage point on the earth, however, was problematic. He needed an exterior vantage point to measure against his calculations of the earth's orbit. He had been making calculations of the Martian orbit as well, so he merely transferred his point of view to Mars:

> Hitherto, the point of view had been from the earth to Mars; now Kepler wanted to follow the earth on its course from a point on the orbit of Mars 'as from a watchtower.' He, so to speak, transposed his eyes to a particular position of Mars' orbit and from there found out directly the relative values of the distances from sun to earth.[73]

The Daemon who, like the Earth Spirit of *Faust*, speaks of the earth to humans does so as an embodiment of the knowledge that it speaks. But this body of knowledge is still fragmentary, and still reliant on the pre- or proto-scientific discourses of "the geographers."

However, as Paxson indicates, this final narrative voice:

> most closely echoes the 'voice' of the outermost narrative frame in the text—the more than two hundred 'discontinuous' though authoritative endnotes sporadically furnished by Kepler through 1630. By certain standards, in fact, the innermost endodiegesis replicates the outermost notational frame *en abyme*, as Dällenbach would assert [. . .] But, just as well, we might follow Derrida by saying that epistemological authority transposes itself with parasitological marginality.[74]

The footnotes are themselves disembodied from the text. They are discontinuous though authoritative, indicative of a moment of modal becoming. The footnotes, which dominate the Daemon's speech, help to expand this section of the text in relation to the surrounding narrative frame. As Mieke Bal notes in her discussion of the "Relations Between Primary and Embedded Texts," "The hierarchical position of the texts is indicated by the fundamental principle of *level*. The relations between narrator's text and actor's text may be of difference in kind and intensity. Quantitative aspect is of influence here: the more sentences frame the actor's text, the stronger is the dependence."[75] Conversely, the more sentences devoted to the interior text, the stronger its independence from the framing narrative. The footnote, as described by Derrida in "This Is Not an Oral Footnote," is part of an already embodied discourse. While "there can be [. . .] footnotes in newspapers," these annotations "find themselves at home" in academic and scientific discourse. In Kepler's *Somnium*, however, these footnotes hint at a discourse in formation. Derrida is quite right to refer to footnotes as "parasites." In this case, the parasite, embodied textually as footnotes, represented in the narrative as the Daemon, is a creature undergoing transformation.

The footnote is not a benign textual feature. The footnote itself has its own history. Thus, while Derrida contends that footnotes belong in academic discourse, this genre as it appears in academic discourse is relatively recent. In *The Footnote*, Anthony Grafton traces the extent to which "the footnote is bound up, in modern life, with the ideology and the technical practices of [the] profession of scholarly endeavor."[76] To this extent, "Even a brief exercise in comparison [of the uses of the footnote] reveals a staggering range of divergent practices under history's apparently stable surface."[77] Derrida identifies, by extension, "The nonbelonging or rigorous, determinable exteriority of the annotation in relation to the principal, primitive text"[78] as a necessary condition for the annotation. However, the qualities that determine this exteriority are themselves contingent on the preconditions governing the discourse.

For Paxson, spatial metaphors are necessary to describe narrative embedding. The charts of the ring compositionists attest to the attractiveness of an endless spatialization of the narrative: the embedded narrative seems to beg for a chart or graph that extends the domain of that narrative, and demonstrates an illusory, metaphoric, and, in the case of the *Somnium*, untenable, symmetry. Indeed, the text itself features a most asymmetric form.[79] As Paxson notes, "the carefully disposed, five-degree sequence of endodiegeses occupying ten folio pages of text (sans the sixth degree, the endnotes) rapidly unwinds like a spring within two sentences."[80] At the end of the narrative, Kepler exposes and, simultaneously, concludes every layer of the narrative:

> When I had reached this point in my dream, a wind arose with the rattle of rain, disturbing my sleep and at the same time wiping out the end of the book acquired at Frankfurt. Therefore, leaving behind the Daemon narrator and her auditors, Duracotus the son with his mother, Fiolxhilde, as they were with their heads covered up, I returned to myself and found my head really covered with the pillow and my body with the blankets.[81]

The end of the narrative serves to reinforce the symmetry of the various levels of the narrative. Each diegetic frame is both opened and closed. The external frame, consisting of Kepler's dream, begins and ends the narrative. This symmetry in terms of narrative levels is not, for Paxson, corroborated by its dureé. The duration of each frame should, in order to merit the symmetry of a narrative, be equal when it is opened and closed. However, the *Somnium* ends abruptly. The end of the book from Frankfurt remains unread. The elements sweep the dream away.

The wind and rain that dissolve the dream are not, however, without significance. At the onset of the narrative, there is no mention of rain. Indeed, Kepler, the astronomer, "went to bed and fell into a very deep sleep" after a night of "watching the stars and the moon."[82] Of course, Kepler couldn't have watched the stars on a stormy night. A storm strong enough to rouse one from sleep, however, let alone from "a very deep sleep" seems quite remarkable, particularly one which arises from nowhere on the Bavarian countryside. But, for Kepler, a dream always heralds a mystical experience, the interpretation of which, and the conditions surrounding it, suggest the language of the divine. The intrusion of the storm marks the dream as a revelation; its power extends beyond unconsciousness.

QUANTIFICATION OF THE WORLD SOUL

The narrative framing of the *Somnium* suggests a view of the natural world that closely parallels the preoccupation with the natural world so evident in the cosmological narratives of the school of Chartres. These narratives reveal a dual interest in conveying philosophical truth and accurate depictions of the natural world. However, while the authors of these narratives might have been interested in depicting the natural world in a more recognizably "accurate" mode than that assumed by their predecessors, they still "raised philosophical issues without providing tools to explore these issues."[83]

While authors such as Alain de Lille and Bernardus Silvestris were concerned with the natural world, "the 'discovery of nature' so crucial to the Burckhardtian view of the twelfth-century Renaissance [. . .] was first and last a rediscovery of texts about nature."[84] Kepler's representation of the natural suggests the influence of these concepts of the natural. For Kepler as well as his predecessors, "the phenomenal world, the *ornatus elementorum* (articulation of the elements), as William and Bernardus Silvestris refer to it, is a tissue of figures and images that must be read like a literary test. From such a standpoint, the philosophy of nature 'involves and embodies a transcendent form of rhetoric.'"[85]

This approach was potentially liberating, but it "proved, in the practice of William and his fellow cosmologists, fatally circumscriptive, for it is here that the limitations of their resources become most plain."[86] Here, Wetherbee distinguishes between the twelfth century cosmologists and their precursors, the grammarians and encyclopedists of late antiquity. In this latter category are included Servius, Macrobius, Calcidius, and Fulgentius. For these writers, "it was axiomatic that the great auctores were repositories of profound philosophical wisdom."[87] However, the twelfth century cosmologists attempted to "ground religious thought in a philosophical understanding of nature and the Liberal Arts."[88]

Critics like R.W. Southern view this as a major reason to discount the significance of these cosmologists. In "Humanism and the School of Chartres," he stipulates that their work was hindered by the limitations of their scientific knowledge.[89] As such, Southern argues that these writers do not differ in any significant manner from earlier writers who interpreted texts to infer facts regarding the natural world. Further, in Christianizing the philosophical concepts of antiquity, allegorists such as William of Conches[90], Abelard[91], and Thierry of

Chartres[92] simply aligned the platonic World Soul with the Christian Holy Spirit, thereby obviating the need for empirical observation.

By no coincidence, Kepler, trained for the clergy and a tremendously committed believer in the presence of divine design in the natural world, presents the Daemon as a version of the World Soul. However, while Kepler follows in this tradition of philosophical writing, the point of view of the narrative obscures the exact origin of the Daemon, complicating the ease with which we can align the Daemon with the Holy Spirit and, correspondingly, the World Soul. Hallyn notes that the origin of the Daemon is of some confusion; we are not certain if the Daemon comes from the moon or the earth. Thus, "The daemon's scientific exposition is acceptable if he is understood from the point of view of a lunary creature, but from another point of view this exposition *also* presents the inverse image of terrestrial science."[93] The section title preceding the speech labels this figure quite clearly as "the Daemon from Levania."[94] Further, the Daemon remarks that "up there we are granted leisure to exercise our minds in accordance with our inclinations. We consult with the daemons of that area and enter into a league," indicating that the moon is a haven for daemons. Still, the daemons also "rush toward the earth with our allied forces [. . .] when mankind sees the sun in eclipse."[95] Thus, the Daemon and his kindred seem to inhabit both Earth and Space.

Such positioning coincides exactly with the placement of celestial beings in the Neoplatonist cosmic model. By including itself within the perspective of earth dwellers, the Daemon ultimately elides a conclusive determination of its origin. Likewise, Kepler refers to the Daemon as Earth Spirit. The parallel to the object of Faust's incantation is uncanny. However, unlike Goethe's Earth Spirit, which merely reveals its own incomprehensibility to Faust, Kepler's Earth Spirit is surprisingly eloquent. Both of these earth spirits have parallels in the allegorical tradition of a conceptual figure representative of the forces linking the natural with the divine.

The twelfth century cosmographers expanded the interpretive possibilities available to the world soul. As Whitman remarks, "they produced a composite figure pointing in two directions [. . .] by consolidating a divine abstraction with a cosmic agent."[96] Thus, the World Soul was an allegorical abstraction which, besides representing the synthesis of Christian doctrine and pagan philosophy, also represented the process by which such a synthesis occurred.[97] To this extent, the conditions under which the world soul is employed play a significant role in determining its meaning: "In its Christian affiliation with divine goodness, this principle could remain otherworldly in its dimensions. In its pagan character as the World Soul, it could be deeply implicated in the world."[98]

Kepler's method, of course, differs significantly from that employed by the Neoplatonists. Thus, while Kepler, like the Neoplatonists, was interested in concepts such as celestial harmony that could be represented in the figure of the World Soul or Earth Spirit, Gerald Holton observes that "Kepler's harmonies reside in the very fact that the relations *are quantitative*, not in some specific simple *form* of the quantitative relations."[99] He goes on to affirm the distinction between Kepler and his predecessors as follows: "It is exactly by this shift which we can now recognize as one point of breakthrough toward the later, modern conception of mathematical law in science."[100] Thus, while Kepler begins from the same philosophical presuppositions informing Neoplatonist cosmographical allegory, he seeks quantitative justification for philosophical and theological concepts.

The *Somnium*, then, must not be viewed as a fanciful allegory placed over Kepler's calculations of planetary orbits. Instead, Kepler questions geocentrism through a logical and mathematically verifiable reversal of the necessary conditions of this model. At the same time, he inaugurates this experiment through an attention to the Neoplatonist conception of cosmic symmetry and proportion. Thus, the Moon, populated by legions of daemons, serves as an anti-sun which exists in the same relation to the earth as the earth exists to the sun. The sun, an allegorical expression of divinity in Neoplatonist cosmology, was the seat of God. Kepler intends for the moon to oppose the sun within his hypothetical account of Levanians who view the moon as the center of the universe. If the daemon-infested moon is the center of the universe, then we may infer that such a model effectively desecrates the vision of the universe as a temple of God Who rules from its center. Even Kepler's use of the name Levania, selected specifically because such a word, derived from Hebrew, "should inspire greater awe and [is] recommended in the occult arts,"[101] suggests a dichotomous relationship between the sun and the moon that echoes classical mythography while at the same time suggesting the sinister implications of an incorrect or misguided perception of the shape of the universe. Indeed, it is only in the Copernican universe that the sun exists at the center. By implication, the geocentric universe, like the selenocentric universe, is misguided, inverted, and, perhaps, diabolically informed.[102]

Mathematics provides Kepler with the necessary tool to explore this theological tangle:

> The investigation of nature becomes an investigation into the thought of God, Whom we can apprehend through the language of mathematics. *Mundus est imago Dei corporea*, just as, on the other hand, *animus*

> *est imago Dei incorporea.* In the end, Kepler's unifying principle for the world of phenomena is not merely the concept of mechanical forces, but God, expressing Himself in mathematical laws.[103]

Holton argues for a standard correlation between Kepler's scientific work and his religious background. Thus, we see Kepler opposed to Descartes, who views the mathematical order of nature as God. Kepler sees the mathematical order of nature as the means by which God expresses His power. Holton offers Kepler's religious training as justification for his perception of Kepler as a theologian working in science, and points out that Kepler often "referred to astronomers as priests of the Deity in the book of nature."[104] The notion of astronomers as priests testifies to the piety inherent in Kepler's work. Holton substantiates his argument further by providing the following statement Kepler made to Herwart von Hohenberg in a letter from December, 1598: "I take religion seriously, I do not play with it."[105] The juxtaposition of these two quotes helps Holton to argue that, for Kepler, nature is religion. Kepler seeks to unravel the mysteries of the book of nature in the same solemn manner of a theologian preparing a Biblical exegesis.

The problem with such an argument is that this is manifestly not the case with the *Somnium.* The exegetical method which Kepler pursues in this text does not correspond to Holton's image of the serious astronomer/ theologian who "does not play" with God's order as it is manifested in scripture or the book of nature. As I have already pointed out, the narrative works on the basis of a fundamental *inversio* of point of view. Kepler exposes the logical inconsistencies of the geocentric model of the solar system through a misrepresentation of a model that truly is geocentric: that is, the orbit of the moon around the earth.

Such *inversio* is a fictionalization, or falsification, of the order of the universe. But we cannot isolate rhetorical methods, regardless of their power and efficacy, from the subjects they are used to represent. Representations of the universe are tied to conceptions of divinity. Copernicanism was so controversial because, if true, it allowed a revised book of nature to rewrite scripture. *Inversio,* however, is the trope of devils; the invocation of a universe where lunar creatures see themselves as the center of the universe implies an overturning of doctrines like salvation through Christ made human, and a belief of humanity created in the image of God. My position here might seem a bit extreme, but similar concerns were raised by Kepler's contemporaries in regards to the people of the New World. On the one hand, writers like Thomas Harriot were content to characterize natives of the New World as examples of prelapsarian innocence. But the question that immediately followed such characterizations involved the status of the souls of the natives. If truly innocent, then why convert them? If they do not know Christ, how can they be considered innocent? Are their barbaric customs and heathen beliefs truly a reflection of man's origins as presented in Genesis, or has the spiritual and material progress of these peoples been stunted through their slavish devotion to deities that are actually demons, actively committed to keeping the natives from accepting the Word of God? The perspective on the natives reflects the perspective on the natural. Likewise, attempts by geometers to disseminate a model of the universe informed by the logic of observation and not the musings found in authoritative sources necessitate a form of argumentation wherein the natural informs the theological.

INVERSION AND THE *NOVA ASTRONOMIA*

Kepler's argumentative methods are perhaps best explained as a result of his academic training. The inversion of meanings evident in the use of the Daemon suggests a sophisticated recourse to the dialectical method. Andreas Planer offers a definition of dialectics as a single method which provides access to all knowledge. Thus, in his *Scientia demonstrandi* (1586), Planer "likens all of knowledge to a building which is approached by only one road, the way or method of demonstration."[106] For Planer, the training of the mind to accomplish the aims of the dialectic is best embodied by Aristotle's *Organon*. Planer seeks to "rid [. . .] all sciences of the errors, opinions, and ignorance shown by so many authors, and he believes that the way to do this is through the proper use of demonstrative method.[107] However, the disparity between the logical aims of a project like Planer's and the actual application of dialectical method became increasingly evident to scholars and educators through the sixteenth century.

Thus, Planer, writing in the late sixteenth century, continues to echo the viewpoint of Philip Melanchthon, who had a direct influence on the curriculum of the University at Tübingen.[108] Planer's description of the aims of the dialectic maintains the credibility of Aristotle's authority because Aristotle's methods emphasize proofs. Still, as Metheun points out, the discrepancy between Aristotle's aims and actual observations became increasingly difficult to ignore. Thus, Melanchthon's "acceptance of Aristotle's authority presumably lies behind [his] unquestioning recourse to Aristotelian cosmology and his rejection of observational evidence which conflicts with that cosmology."[109]

The messages provided through observation of the natural world, however, also hint at a need to readdress Aristotle's favored methods as well. Dialectics, which "makes it possible to go beyond 'common appearances' to what is hidden, allowing the essence of natural things to be ascertained by contemplation"[110] still connotes a divinely guided pursuit of knowledge, rather than a system of proofs rigorously aligned with observation, and not interpretation.

The contrast between methods here becomes clearly evident when Aristotle's philosophical aims clearly contradict Aristotle's recorded observations. Michael Maestlin's observations of the 1572 nova and two comets highlight this problem. In his then-controversial findings, based on his measurements of the parallax of such phenomena, "Maestlin concludes that they are all supralunar, rather than sublunar, and thus he contradicts the teachings of Aristotle that comets are sublunar and that no change can occur in the supralunar region."[111] While "Maestlin is convinced that an accurate understanding of God's creation will lead to a more precise knowledge of God and of God's intentions for the world,"[112] he places observation above appeal to traditional authorities. The precision of his observations, derived through geometrical proofs, allows him "to draw conclusions, the truth and certainty of which are to be rated higher than the authority of the opinions of Aristotle, Pliny, and other ancient philosophers."[113]

I highlight this conflict with various approaches to knowledge as an example of the various possibilities available to Kepler as a scientist. At the same time, I think that these problems surrounding the presentation of knowledge become manifested in the form of the Daemon. The University of Tübingen was the source of heated discussion on method and presentation. Such views were, likewise, reflected in the curriculum of the school and the publications of its instructors. For example, Maestlin's conclusions regarding the nova of 1572 do more than discredit the authority of Aristotle's conception of the natural world. Instead, a text such as Maestlin's *Demonstratio astronomica loci stellae novae* argues that the existence of a new star "represents a change in the heavens, not simply above the moon, but in the sphere of the stars, previously assumed to be perfect and immutable."[114] This does not only damage a particular model of the universe: the force of observation does more than discredit the Neoplatonist model of the universe which had, by the sixteenth century, fused classical philosophy and Christian theology into a set of tools used for describing the natural world. Instead, it casts doubt on the pairing of perfection and immutability so central to what is essentially a geometrical model of the universe.

The scholars of the University of Tübingen and their students were committed to expressions of divinity in the natural world. Indeed, Kepler's early *Mysterium Cosmographicum* builds from the notion of the geometric perfection of the solar system. Unlike the Neoplatonist model, the *nova astronomia* relies principally on geometrical proofs and mathematical calculations in order to support observation. However, these proofs and calculations are still used in order to verify the perfection of divine design. Thus, as Kepler notes in the dedicatory epistle to the *Mysterium Cosmographicum*, "For the more rightly we understand the nature and scope of what our God has founded, the more devout our spirit will become."[115]

The goals of the *Mysterium Cosmographicum* are not, of course, the same as those of Kepler's later works. Still, Kepler's debt to his training at Tübingen is evident throughout his career, leading him to "a theological mathematics, that is, to the precise observation and interpretation of the heavens in God's name."[116] At the same time, this conflict between perfectibility and immutability and the connection of these factors to method suggests some things about the character of the Daemon in the *Somnium*. Regardless of the etymological roots of the word daemon, the name still carries a connotation of diabolism, of a reversal or overturning of holiness. Furthermore, while Kepler playfully likens representations of astronomical observation to the conventions of demonic summoning, his own views on the essentially theological aim of astronomy beg the question why he uses a daemon as a mouthpiece for Copernicanism. The language of the Daemon is most certainly a recognizably scientific language. Unlike the burbling curses and shrieking hisses of the demons populating Bosch's paintings, Kepler's Daemon makes learned statements, noting, for instance, that "The intersections of the equatorial and zodiacal circles create four cardinal points, like our equinoxes and solstices" and "For they indicate the longitude of places with reference to their motionless Volva, and the latitude with reference both to Volva and to the poles, whereas for longitudes we have nothing but that most lowly and barely perceptible declination of the magnet."[117]

But we cannot merely equate diabolism with barbarism, and thereby conclude that Kepler's Daemon has nothing in common with Christian representations of the Enemy and His minions. Indeed, while some demons are represented as sinful urges personified in monstrous forms, others can also be depicted as highly educated figures who use their learning in order to tempt and torment. While Goethe's Mephistopheles is surely the most recognizable prototype of such a demon, the Biblical representation of Jesus' temptation in the forest, or even of Satan as diabolical advocate in *Job*, hint at the complexities of the persona of absolute evil.

The Daemon is, then, demonic in the Christian sense. It is the devil of Job, perhaps, playing the role of the inquisitor in the court of God. It is a devil who, by "Roaming through the earth and going to and fro in it,"[118] knows the natural in ways unavailable to the angels assembled in heaven.

Indeed, the Daemon, though undescribed, and undescribable by Duracotus, who lies huddled under a sheet, perhaps resembles the foreboding Devil of the Brixen altarpiece. The demon of Christian imagination, it is adorned with leathery wings, fangs, and horns. But also, it appears as a teacher, proffering a book to its unwitting students. Still, the invocation of the Daemon implies a complex dialogue. As Michel Serres notes in *Angels: A Modern Myth*:

> By a juridicial logic, we choose to depict as devils those who cause us suffering, and who enjoy such a power that they would win a trial against us from the very moment that we publicly brought a plea against them. [. . .] But has anyone ever really been scared of this skinny beast, this poor horned devil with eyes in his bottom, this victim of our cruel weakness?[119]

At the same time, perhaps it is wrong to use such an image to evoke the Daemon. Kepler, in eliding description, and aligning the voice of the Daemon with the rolling thunder of a storm, connects the Daemon to the forces of nature that it represents. In doing so, Kepler presents us with a personage that is angelic in Serres' sense of the angel as a manifestation of "the beauty of the world."[120] This is not, however, an irresolvable contradiction. The representation of the Daemon is, moreover, closely paired with Kepler's attempt to resolve the competing means of representing nature that were most readily available. The angel in one of these modes of discourse becomes a demon in the next.

While the Daemon is a force of nature, there are, for Kepler, three different levels or ways of viewing nature. The first views the *auctoritas* as the ultimate repository for information about the natural world. The second such approach attempts to synthesize Neoplatonist philosophies and observations of the natural world. This attempt at synthesis, as Wetherbee argues, is the ultimate limitation of this approach. Thus, while the twelfth century cosmographers attempt to discard this approach, they are still indebted to the *auctoritas* to the extent that they can never get to a science based primarily on calculations. While they might affirm Aristotle's view that we must use observation in order to examine the natural world, they are still too likely to see Aristotle's own observations, many of which are very inaccurate, as admissible because of their origin. This approach would also reflect the degree to which the writings of classical authorities were Christianized: even if they were pagan, much of their work was interpreted in such a way as to highlight the Christian leanings or potential.

Kepler, though, is faced with observations and data that clearly defy the laws of tradition. The third approach follows from this contradiction: nature is that which is verified through empirical observation and mathematics. Kepler, however, does not entirely embrace this approach, though he senses its validity. He still employs data as a way to justify the theological concepts underlying Neoplatonism. For a scientist so heavily influenced by the mysticism of Neoplatonism, Kepler employs the Daemon as a way of questioning the validity of his findings. While Kepler is clearly convinced of Copernicanism, he sets out to justify this position in relation to previous conceptions of astronomy. He sees this as necessary for the continuation of astronomy as a viable and productive science and explanation of the theological question of the universe. The Daemon is, then, a figure that provides Kepler with a composite of references and polymorphically potential interpretations of meaning. This range of interpretations prevents us from reading the Daemon as a closed figure. The multiple meanings of the Daemon signify, like the universe it describes, the limitations of signification itself. Thus, the form of the dream allegory, previously a form with a claim to truth derived from mystical inspiration, becomes, like the universe itself, a place of uncertainty, a form made formless.

AFTERWORD

Franz Kafka remarked once that writing is the reward for service to the devil. I've thought about this quote frequently while writing this book. While Kafka is speaking in a very different cultural context, I wonder if his statement somehow reflects Kepler's quandary. In fact, Kepler's dream is very much a *written* artifact; for much of his life, Kepler's fingers were stained with ink from this text. But Kepler's continuous writing of the *Somnium* never absolved him from the sin of discovery, from the creation of a new universe made through observation, conjecture, and imagination. Instead, Kepler, writing in an age of religious turmoil, his fortunes shaped by this turmoil, finds himself attracted to an even more dangerous form of turmoil: the polysemous book of the universe.

Notes

NOTES TO CHAPTER ONE

1. *The Travels of Sir John Mandeville,* trans. C.W.R.D. Moseley (London: Penguin, 1983), 183.
2. Ibid.
3. Ibid., 184.
4. Ibid., 185.
5. Ibid.
6. Ibid.
7. Marjorie Nicolson, *Science and Imagination* (Hamden: Archon Books, 1976), 41.
8. Michel Serres, *Angels: A Modern Myth,* trans. Francis Cowper (Paris: Flammarion, 1995). In particular, Serres' poetic evocation of the often-overlooked alliance of the supernatural and the technological in this book suggests, I would argue, a sentiment more in keeping with the influence of theology on the development of scientific language.
9. Silvestris, 5.
10. Bruce Stephenson, *Kepler's Physical Astronomy* (Princeton: Princeton University Press, 1994), 202.
11. Robert M. Jordan, *Chaucer and the Shape of Creation: The Aesthetic Possibilities of Inorganic Structure* (Cambridge: Harvard University Press, 1967), 15.
12. Ibid., 20.
13. The mathematic formulations of the Ptolemaic world system do feature numbers and proportions based on ratios. However, the mathematic basis of the system stems from interpretations of numbers as philosophical concepts. As Stahl reminds us in his introduction to *The Commentary on the Dream of Scipio,* "the monad, which Pythagoreans derived from menein 'to remain' [. . .], refers to the unchangeable and imperishable, the One, Mind, and Soul [. . .] and also refers to the celestial sphere, which [. . .] shares in the divinity of the Soul" (Macrobius 103). Thus, this section of the macrocosmos is aligned with the number one because both are unchangeable.

Likewise, one is a barrier (for whole numbers at least) between changeable numbers and nothingness. Two, the dyad, like "the errant spheres of the planets and the sun and moon" (Macrobius 103), represents change and movement and is bound by the non-movement of the monad. Macrobius, *Commentary on the Dream of Scipio by Macrobius,* trans. and ed. William Harris Stahl (New York: Columbia University Press, 1990).

14. The etymology of the Greek term for the original philosophers, physiologoi, suggests a similar journey: "These first philosophers investigated physis (nature), trying to discover its arche (origin) by means of reason (logos)" (Koyama viii).
15. Macrobius, 103.
16. Plato, *Plato's Cosmology,* trans. F.M. Cornford (New York: Humanities Press, 1952), 30d.
17. Rosen, Edward. *Three Imperial Mathematicians: Kepler Trapped Between Tycho Brahe and Ursus* (New York: Abaris, 1986), 13 and 210.
18. See Rosen, 210. Evidence for the dissolution of planetary spheres in the Tychonic system rests more on cognitive illusion than nuanced argument. Howard Margolis exposes the fallacious argument that the apparent collision of Mars and the sun leads automatically to the abandonment of celestial spheres, using Brahe's own diagram to demonstrate that "Mars could no more collide with that mathematical locus than a ship could collide with the equator." Margolis went so far as to contact renowned historian of science, Owen Gingerich, to convince him of this persistent illusion. Indeed, in "The Tycho Illusion," Gingerich defends Margolis' claims against David Topper's refutation, and writes that Topper "has apparently failed to make the cuts correctly."
19. Patrick Moore, *The Great Astronomical Revolution: 1534–1687 and the Space Age Epilogue* (Chichester: Albion, 1994), 104.
20. Hans Robert Jauss, *Question and Answer: Forms of Dialogic Understanding,* trans. and ed. Michael Hays (Minneapolis: University of Minnesota Press, 1989), 72.
21. Fernand Hallyn, "Preface," in *Metaphor and Analogy in the Sciences,* ed. Fernand Hallyn (Dordrecht: Kluwer Academic Publishers, 2000), vii.
22. Jon Whitman, *Interpretation and Allegory: Antiquity to the Modern Period* (Leiden: Brill, 2000), 20.
23. Peter Szondi, *Introduction to Literary Hermeneutics,* trans. Martha Woodmansee (Cambridge: Cambridge University Press, 1995), 114.
24. Angus Fletcher, *Allegory: The Theory of a Symbolic Mode* (Ithaca: Cornell University Press, 1964), 305.
25. Jeffrey Jerome Cohen, "Monster Culture (Seven Theses)," in *Monster Theory,* ed. Jeffrey Jerome Cohen (Minneapolis: University of Minnesota Press, 1996), 6.
26. Ibid.

27. Theresa M. Kelley, *Reinventing Allegory* (Cambridge: Cambridge University Press, 1997), 13.

NOTES TO CHAPTER TWO

1. Wolfgang Kluxen, "Nature in the Ethics of the Middle Ages," in *Nature in Medieval Thought: Some Approaches East and West,* ed. Chumaru Koyama (Brill: Leiden, 2000), 35.
2. Arthur Koestler, *The Watershed: A Biography of Johannes Kepler* (New York: Macmillan, 1960), 149.
3. Carlos Steel, "Nature as Object of Science: On the Medieval Contribution to a Science of Nature," in *Nature in Medieval Thought: Some Approaches East and West,* ed. Chumaru Koyama (Brill: Leiden, 2000), 125.
4. C.S. Lewis, *The Discarded Image* (Cambridge: Cambridge University Press, 1964), 99.
5. By extension, Fredric Jameson's positioning of the Frank Gehry house as an exprcssion of postmodern allegory, as well as the shape of the postmodern imagination, precisely because of its openness, the extent to which it allows contemplation of an infinite universe made of chain-link and aluminum siding, makes us question the other end of the spectrum of the Romantic sublime: does the sublime currently exist, or have we replaced the myth of infinite nature with a cosmological model built from the same materials as the Gehry house, the omnipresent materials of eternal economic expansion?
6. Alexandre Koyré, *Galileo Studies,* trans. John Mepham (Atlantic Highlands: Humanities Press, 1978), 5.
7. Lewis, 100.
8. Ibid.
9. John M. Steadman, *Nature Into Myth: Medieval and Renaissance Moral Symbols* (Pittsburgh: Duquesne University Press, 1979), 3.
10. Piero Boitani, *The Tragic and the Sublime in Medieval Literature* (Cambridge: Cambridge University Press, 1989), 252.
11. Max Caspar, *Kepler,* trans. C. Doris Hellman (New York: Dover, 1993), 154.
12. J.V. Field, *Kepler's Geometrical Cosmology* (London: Athlone Press, 1988), 19.
13. Caspar, 18.
14. Ibid.
15. Field, 18.
16. Simon Brittan, *Poetry, Symbol, and Allegory: Interpreting Metaphorical Language from Plato to the Present* (Charlottesville: University of Virginia Press, 2003), 54.
17. Ibid., 53.

18. Field, 18.
19. Bruce Clarke, *Energy Forms: Allegory and Science in the Era of Classical Thermodynamics* (Ann Arbor: University of Michigan Press, 2001), 18.
20. Gay Clifford, *The Transformations of Allegory* (London: Routledge, 1974), 14.
21. Ibid., 11.
22. Ibid., 15.
23. Clarke, *Energy*, 35.
24. Examples taken from Book I of *The Faerie Queene.*
25. Clifford, 14.
26. Brian Stock, *Myth and Science in the Twelfth Century: A Study of Bernard Silvester* (Princeton: Princeton University Press, 1972), 150.
27. See "Poetry, Revisionism, Repression" in Harold Bloom's *Poetry and Repression: Revisionism from Blake to Stevens* (Yale University Press, 1976) for a complete exploration of this topic.
28. Piet Schrijvers, from the Universiteit Leiden, presented his talk on "Longinus and Quintilianus: The Classical Sublime" at *Shadows of the Sublime: History of a Concept, the Third Gent Conference on Literary Theory*, held at Universiteit Gent in October, 2002.
29. See the *Odes,* 23–25.
30. Bloom, 339.
31. Clifford, 3.
32. Ibid.
33. Ibid.
34. Jim Paxson, *The Poetics of Personification* (Cambridge: Cambridge University Press, 1994), 118.
35. Mieke Bal, *Narratology: Introduction to the Theory of Narrative,* trans. Christine van Boheemen (Toronto: University of Toronto Press, 1994), 93.
36. Ibid.
37. Ibid.
38. Francis Xavier Newman, *Somnium: Medieval Theories of Dreaming and the Form of Vision Poetry* (PhD diss., Princeton University, 1963), 5.
39. Ibid., 6.
40. Salomon Resnik, *La mise en scène du rêve,* trans. Gabriella Mazzini (Paris: Payot, 1984), 10. Translation mine.
41. Artemidorus Daldianus, *The Interpretation of Dreams: The Oneirocritica of Artemidorus,* trans. and ed. Robert White (Torrance: Original Books, 1990), 5.

NOTES TO CHAPTER THREE

1. Although, despite the transcendence of the sign so obviously in evidence in the *Paradiso,* Dante's evocation of the Divine is never entirely divorced from more earthly concerns. In the *Letter to Cangrande,* Dante goes so far as to liken divinity, as represented in Dante's work, to his patron: "And I could

find nothing more fitting to even your great eminence than that sublime canticle of the Comedy which is adorned with the title of Paradise. And so, dedicated to your self, and with this letter serving as its introduction, I inscribe it to you, offer it to you, and, in short, commend it to you" (Dante, *Criticism* 97). Of course, the *amplificatio* evident in this inscription is not a device specific to Dante, and was a common strategy for securing favor and prestige. The dedicatory epistle of the *Faerie Queene,* for example, likewise elevates a patron to the position of a deity.

2. Boitani, 223.
3. Ibid.
4. Frederic Jameson, *The Prison-House of Language: A Critical Account of Structuralism and Russian Formalism* (Princeton: Princeton University Press, 1972), 88.
5. Ibid.
6. A.C. Crombie, *Medieval and Early Modern Science,* vol. 1 (New York: Doubleday, 1959), 75.
7. Tzvetan Todorov, *The Poetics of Prose,* trans. Richard Howard (Ithaca: Cornell University Press, 1977), 138.
8. Italo Calvino, *The Castle of Crossed Destinies,* trans. William Weaver (New York: Harvest, 1979), 39.
9. Rosemond Tuve, *Allegorical Imagery: Some Mediaeval Books and Their Posterity* (Princeton: Princeton University Pres, 1966), 21.
10. Paul Ricoeur, "Energetics and Hermeneutics in *The Interpretation of Dreams,*" in *Modern Critical Interpretations: Sigmund Freud's The Interpretation of Dreams,* ed. Harold Bloom (New York: Chelsea House, 1986), 73.
11. Stephen Prickett, *Narrative, Religion and Science: Fundamentalism Versus Irony, 1700–1999* (Cambridge: Cambridge University Press, 2002), 97.
12. Robert M. Jordan, *Chaucer and the Shape of Creation: The Aesthetic Possibilities of Inorganic Structure* (Cambridge: Harvard University Press, 1967), 29.
13. Augustine, *On Christian Doctrine,* trans. D.W. Robertson (Indianapolis: Bobbs-Merrill, 1958), 34.
14. Eric Jager, *The Tempter's Voice: Language and the Fall in Medieval Literature* (Ithaca: Cornell University Press, 1993), 58.
15. Ibid., 101.
16. Eugene Vance, *Mervelous Signals: Poetics and Sign Theory in the Middle Ages* (Lincoln: University of Nebraska Press, 1986), 36.
17. Ibid.
18. Ibid., 48.
19. Jordan, 26.
20. Such a relation between language and representation did not animate all medieval genres. Part of this connection between signification and divinity stems from the medieval cosmological allegory as a mode concerned with the transcendent. The fabliaux would, perhaps, be the genre in direct

opposition to this philosophy of language. The fabliaux, built from "displacements, condensations, and substitutions [. . .] disrupts the assumption of a 'natural' relation between language and meaning and, at the same time, serves as a screen for the fact that such a relation never existed in the first place" (Bloch 128). The undercurrent of the fabliaux reminds us of the extent to which the narratives I am concerned with in this project attempted to formulate totalizing sign systems. But this attitude toward expression was not characteristic of all literary texts at the time. I am reminded of Milan Kundera's *The Joke* as an expression of two simultaneously held but very different attitudes toward the meanings of language.

21. Martine Dulaey, *Le Rêve dans la vie et la pensée de Saint Augustin* (Paris: Études Augustiniennes, 1973), 165. Translation mine.
22. Vance, 47.
23. Ibid.
24. Qtd. in Vance, 48.
25. For medieval and early modern thought, both music and mathematics were totalizing systems: "music comprehends all temporal movements, and geometry comprehends all spatial forms" (Jordan 21). However, arithmetic underlies both time and space. Thus, arithmetic "provid[es] the absolute principle to which both refer" (Jordan 21).
26. Ibid., 35.
27. Jacques Derrida, *Of Grammatology,* trans. Gayatri C. Spivak (Baltimore: Johns Hopkins University Press, 1976), 14.
28. Macrobius, 86.
29. Ibid., 86.
30. Plato's *Timaeus* in particular exerted a considerable influence on Neoplatonists intent on fusing Plato's account of the creation with Christian doctrine. As Raymond Klibansky notes in *The Continuity of the Platonic Tradition During the Middle Ages,* the *Timaeus* "was studied and quoted throughout the Middle Ages, and there was hardly a mediaeval library of any standing which had not a copy of Chalcidius' version and sometimes also a copy of the fragment translated by Cicero" (Klibansky 28).
31. Macrobius, 86.
32. In this example of philosophical concepts linked to physical forms, nature may serve [to] mirror [. . .] transcendental realities." As a result, "the true objects of contemplation are ideas; the concrete objects of nature serve primarily as figurative lenses for apprehending them" (Steadman 3).
33. Copeland, 43.
34. Such concerns were further exacerbated by plagues, sickness, and disaster. See George Holmes' *The Later Middle Ages, 1272–1485* pp. 136–138 for a discussion of sickness and artistic production in England.
35. Marc Cogan, *The Design in the Wax: The Structure of the Divine Comedy and Its Meaning* (Notre Dame: University of Notre Dame Press, 1999), 176.

36. Jon Whitman, *Allegory: The Dynamics of an Ancient and Medieval Technique* (Oxford: Clarendon Press, 1987), 52.
37. Ibid.
38. Ibid., 53.
39. Ibid.
40. Ibid.
41. Ibid.
42. In his introduction to Matthew of Vendôme's *Ars Versificatoria,* Roger Parr remarks that "by the eleventh century the flowering of the language arts is reflected in the diversity of manual titles: *Ars Praedicandi* (thematic sermon), *Ars Dictaminis* (letter writing), *Ars Rithmica* (rhythmic composition), *Ars Notaria* (legal documentation), *Ars Disputatio* (argumentation), *Rhetorica Ecclesiastica* (church rhetoric), *De Schematibus et Tropis* (figures and tropes), *Ars Metrica, Ars Poetria, Ars Versificatoria* (versification). The flood of handbooks indicates clearly the pragmatic exploitation of the devices of style of classical rhetoric" (Vendôme 3). These diverse discussions of types of communication emphasize the dominance of rhetoric as the basis of education.
43. Jordan sees this as a fundamental cause for the late-medieval interest in rhetoric:

    > The humanists' emphasis upon the verbal surface of poetry, that is, upon the composition of the 'veil,' was congenial to a rhetorical approach to the art of poetry. The revival of Ciceronian and Horatian precepts in the popular rhetorical handbooks, such as the *Poetria nova* of Geoffrey of Vinsauf and the *Ars versificatoria* of Matthew of Vendôme, substantiated and enlarged the concept of poetry as a quantitative or 'inorganic' art, that is, an art concerned with the management and disposition of the fixed elements constituting a preconceived whole. (Jordan 42)

    Such guidebooks allowed rhetoricians to articulate, in great detail, the parts composing the range of rhetorical possibilities. Such an encyclopedic documentation promised a greater ability to fathom meanings concealed by the veil of the text.
44. Whitman, 54.
45. See Parr's introduction to Vendôme, in particular pp. 3–6.
46. Whitman, 54.
47. Alain, *Complaint,* 117.

NOTES TO CHAPTER FOUR

1. The date of composition cannot be accurately determined for *The House of Fame.* As George K. Anderson comments in *Old and Middle English Literature from the Beginnings to 1485,* "In deciding on the date and purpose

of *The House of Fame*, we are confronted with a difficult, virtually unanswerable question. The poem is written in octosyllabic couplets, indicative of Chaucer's early works; but its author is obligated to Roman and Italian writers, notably to Virgil and Dante, and the tone of the poem is the tone of the mature Chaucer. By general consent it is placed before *Troilus and Criseyde*, and therefore somewhere between Chaucer's first Italian visit and 1382" (Anderson 236-37).

2. Line 527.
3. Line 541.
4. Line 592.
5. Lines 586-87.
6. Ovid, *Metamorphoses*, trans. Horace Gregory (New York: Mentor, 1960), 356-57.
7. Both of these objects, transcribed into the stars, could have also had religious significance. See Barry B. Powell's *A Short Introduction to Classical Myth*, pp. 105-111, for a discussion of the myths and religion surrounding Dionysus (and his wife Ariadne) and Orpheus.
8. Lines 661-63.
9. See Kathryn L. Lynch's "Partitioned Fictions: The Meaning and Importance of Walls in Chaucer's Poetry" for a discussion of Chaucer's "enclosed fictions" (Lynch 107). Lynch argues that enclosed settings characterize Chaucer's dream visions.
10. While I am classifying Chaucer's *The House of Fame* as allegorical, Maureen Quilligan contends that it is "not quite allegorical because of its parodic form" (Quilligan 247).
11. Thomas Bulfinch, *Bulfinch's Mythology* (New York: Gramercy Books, 1979), 889.
12. Peter Dronke, *Sources of Inspiration: Studies in Literary Transformations, 400-1500* (Rome: Edizioni di Storia e Letteratura, 1997), 177.
13. Jauss, 9.
14. Alain, *Anticlaudianus*, 120.
15. Ibid.
16. Friedrich Heer, *The Medieval World: Europe 1100-1350*, trans. Janet Sondheimer (New York: Mentor, 1962), 306-07.
17. Beal, *Religion and its Monsters*, 15.
18. Silvestris, 123.
19. Robert R. Edwards, *The Dream of Chaucer: Representation and Reflection in the Early Narratives* (Durham: Duke University Press, 1989), 118.
20. Bulfinch, 127.
21. Pierre Brunel, *Histoire de la Littérature Française du Moyen Âge au XVIII siècle* (Paris: Bordas, 2001), 47. Translation mine.
22. See Robert White's Introduction to *The Interpretation of Dreams: The Oneirocritica of Artemidorus*, 5.

23. Steven F. Kruger, *Dreaming in the Middle Ages* (Cambridge: Cambridge University Press, 1992), 7.
24. Ibid., 8.
25. Lynn Thorndike, *A History of Magic and Experimental Science: During the First Thirteen Centuries of Our Era, Vol. II* (New York: Columbia University Press, 1957), 291.
26. Kruger, 8.
27. Thorndike, *History*, 296.
28. Ibid.
29. Qtd. in Kruger, 9.
30. Ibid., 9-10.
31. François Berriot, *Exposicions et significacions des songes et Les songes Daniel* (Genève: Librairie Droz, 1989), 32. Translations my own unless otherwise indicated.
32. Ibid., 9.
33. Thorndike, *History*, 295.
34. Ibid., 296.
35. Jorge Luis Borges, *Libro de Sueños* (Madrid: Alianza Editorial, 2000), 7. Translations my own unless otherwise indicated.
36. Ibid., 7.
37. Macrobius, 87 n.
38. Ibid., 88.
39. The *Oneirocriticon*, or Ὀνειροκρυτυκά, was widely read and translated by medieval audiences. An early known version was printed by the Aldine press at Venice in 1518; a Latin translation by Cornarius appeared at Basel, 1539; it was published in both Latin and Greek by N. Rigaltius at Paris, 1603. (Thorndike, *History* 291)
40. Artemidorus, 16.
41. Thorndike, *History*, 291.
42. As Kruger points out, these claims to Biblical authority were not validated by canon law. Indeed, "the most influential medieval collection of canon law, Gratian's *Decretum*, condemns more or less explicitly" any kind of dream book. Furthermore, "in the *Decretum*, Gratian similarly moves to negate the claims of authority attached to the extremely popular *Somniale Danielis*" (Kruger 12). In the *Decretum*, Gratian likewise unambiguously denies the attribution of dream books to Daniel: "somnialia scripta, et falso in Danielis nomine intitulata" [dreambooks written down and entitled with the false name of Daniel] (qtd. in Kruger 13).
43. Thorndike, *History*, 291.
44. Macrobius, 92.
45. Ibid.
46. Ibid.
47. Artemidorus, 15.

48. Qtd. in Kruger, 45.
49. Macrobius, 88.
50. Ibid., 89.
51. Ibid., 89-90.
52. Kruger, 22.
53. Macrobius, 89.
54. Kruger, 22.
55. In *Somnium: Medieval Theories of Dreaming and the Form of Vision Poetry*, Francis Xavier Newman designates true and false dreams as 'divine' and 'diagnostic.' The diagnostic dream originates in the body so that "in their technique of interpreting dream images, the doctors are very similar to the diviners" (Newton 45). Newman points out that, with this distinction of causes, dreams become a medical ailment. He notes that, unlike Artemidorus, Hippocrates "restricts all his interpretations [of dreams] to medical matters" (Newman 45).
56. Qtd. in Kruger, 20.
57. Newman, 68.
58. Kruger, 18.
59. Newman points out that Macrobius, along with Iamblichus and Calcidius, "emphasize that the highest kind of dream confers *gnosis*. This means a kind of wisdom, which is both esoteric, transcending man's ordinary grasp, and moral, initiating man into a way of life that brings him closer to divinity" (Newman 74). While Macrobius used a pagan frame of reference to express such concepts, this tendency toward valuing mystical experience would be of key importance to the continued usefulness of these categories for Christian commentators.
60. Macrobius, 90.
61. Ibid.
62. Ibid.
63. Ibid.
64. Joyce writes:

> He turned to the flyleaf of the geography and read what he had written there: himself, his name and where he was.
>
> *Stephen Dedalus*
> *Class of Elements*
> *Clongowes Wood College*
> *Sallins*
> *County Kildare*
> *Ireland*
> *Europe*
> *The World*
> *The Universe*. (Joyce 27)

65. Macrobius, 90.

66. Rosemond Tuve comments on a similar organizational principle found in Macrobius' categorization and personification of the virtues: "His implication is that these elements, which follow from, or come of, or show, or declare, the great quiddity he is dividing, are aspects or faces in which the virtue is made manifest. At any rate, this is the understanding maintained with a curious clarity even in popular treatments using Macrobius. This is a typical mediaeval way view such a matter, and it is figuratively fertile. It makes allegory a natural mode in which to portray the visible action of such universals, which were difficult to grasp and hold in a definition" (Tuve 70). While Tuve refers to the virtues, we can likewise consider this attitude toward classification in the context of dream categories. They themselves are not personified; however, the overlap of dream categories is suggestive of a similar logic regarding the presentation of dreams and virtues as concepts.
67. Macrobius, 90.
68. Kruger, 23.
69. Ibid.
70. Ibid.
71. Qtd. in Kruger, 23.
72. Macrobius, 85.
73. Whitman, *Interpretation*, 52.
74. Macrobius, 92.
75. The Neoplatonism of the late medieval commentators was an attempt to reconcile the pre-Christian conception of all earthly phenomena. Thus, "these writers consciously refused to depict dream experience as unvarious" (Kruger 20). Despite "Plato's scattered comments on dreaming" (Kruger 26), the Neoplatonists succeeded in constructing "a simultaneously dual and hierarchical structure central to the classification of dreams" (Kruger 32).
76. Ludwig Schrader, "Joachim Du Bellays 'Songe' (1558)—eine Vision des Untergangs" in *Traum und Träumen*, ed. Rudolf Hiestand (Düsseldorf: Droste Verlag, 1994), 89. Translation mine.
77. Artemidorus, 208.
78. Ralph B. Crum, *Scientific Thought in Poetry* (New York: AMS Press, Inc., 1986), 39.
79. Newman, 52.
80. Lucretius' madness, however, has long been a subject of discussion. Cicero, among others, has commented on Lucretius' madness. As Mayotte Bollack observes in *La raison de Lucrèce: Constitution d'une poétique philosophique avec un essai d'interprétation de la critique lucrétienne*, "la folie est poétique et amoureuse à la fois [. . .] mais le travail de l'artiste, bien qu'inspiré par elle, ne coïncide pas avec le délire. Lucrèce ne s'abandonne pas à cette inspiration; il écrit contre elle, en elle, si bien qu'elle est traitée à la fois comme un moyen et comme une contrainte" [the madness is poetic and amorous at the same time but the work of the artist, while inspired by it, does not

coincide with madness. Lucretius did not abandon himself to this inspiration; he wrote against it, in it, and thus it should be viewed at the same time as enabling and as a constraint] (Bollack 76). Translation mine.

81. Macrobius, 85.
82. Ibid., 86.
83. See, for instance, Abelard's *Epitome theologiae Christianae*.
84. Peter Machamer, "The Nature of Metaphor and Scientific Description" in *Metaphor and Analogy in the Sciences*, ed. Fernand Hallyn (Dordrecht: Kluwer Academic Publishers, 2000), 89.
85. Ibid., 39.
86. Ibid., 39-40.
87. Ibid., 40.
88. Ibid., 52.
89. See my further discussion of this problem in "The Limits of Language as a Celestial Vehicle." This problem is closely linked to the development of the practices of medieval hermeneutics and exerts a considerable influence on medieval and early modern allegory:

    > Though strongly secular in their persuasion, the new humanists of the late Middle Ages shared with the Church Fathers the understanding that poetic utterance is literally untrue. St. Jerome, for example, asserted that the poet's business, like the prophets, was not to speak plainly but to speak in such terms as only the initiated could understand. It was understood that the obscurity of the surface and the need to penetrate it enhanced the value of the hidden truths and preserved them from vulgarization. In adopting this symbolic conception of poetry the humanists were borrowing the prestige of traditional spiritual exegesis. (Jordan 7)

90. Although, the animal fable was an important literary form: "la littérature latine médiévale n'ignorait pas les contes d'animaux" [Latin medieval literature did not ignore the animal fable] (Brunel 43). Furthermore, while we can not, of course, read biblical texts as animal fables, the representation of animals as the embodiment of instincts and virtues, animal fables "continuaient la tradition antique de Phèdre et d'Ésope" [continued the classical tradition of Phaedre and Aesop] (Brunel 47) and exerted a considerable influence on medieval representation.
91. Peter Schmidt, *Het Lam Gods, Gent* (Gent: Ludion, 2001), 12. Translation mine.
92. However, Van Eyck was not only concerned with the allegorical and otherworldly. In the late middle ages, "an interest in the visible, empirically definable world insured that naturalism interpenetrated the classical revival in numerous ways" (Stock 6). Thus, in the visual arts, "the saints and the heroes of antiquity are not eternal archetypes, models of wisdom and of action, but begin to resemble the citizens of medieval towns" (Stock 6).

93. Macrobius, 82.
94. Ibid.
95. Ibid., 84.
96. Ibid.
97. Ibid., 85.
98. Ibid.
99. Ibid.
100. Ibid.
101. Ibid.
102. Ibid.

NOTES TO CHAPTER FIVE

1. François Berriot attests to the popularity of this genre, noting that "A la Renaissance, le Livre de Daniel [. . .] après avoir longtemps circulé sous forme manuscrite, sont reproduits par l'imprimerie en latin, français, italien, anglais, allemand, et deviennent [une des] les deux grands classiques de ce genre 'scientifique' extrêmement en vogue en Occident et dans le bassin méditerranéen depuis la fin du Moyen Age jusqu'au temps du baroque"[In the Renaissance, the Book of Daniel [. . .] after being long circulated in manuscript form, was reprinted in Latin, French, Italian, English, German, and became [one of] the two great classics of this "scientific" genre which was extremely fashionable in the west and in the mediterranean basin from the end of the Middle Ages until the baroque period.] (Berriot, *Spiritualités* 393). Translation mine.
2. Dante's treatment of and recuperation of Virgil in the *Commedia* is perhaps the best known example of this process.
3. John of Salisbury was particularly concerned with the attitude toward citations of authority: "the rule of the *auctores* was shaken in the twelfth century not only by the triumphant progress of dialectics (now called logic) but equally by the revolt of youth against the traditional school curriculum. John of Salisbury already has to resist the new trend in his *Metalogicon* and his *Entheticus.* This trend, he complains, scorns the *auctores,* grammar, and rhetoric" (Curtius 53).
4. John of Salisbury, *Ioannis Saresberiensis episcope carnotensis Policratici,* 2 vols, ed. Clemens C.I. Webb (Frankfurt: Minerva G.M.B.H., 1909), I 9.
5. Curtius, 236.
6. Qtd. in Curtius, 77.
7. Fletcher, 108.
8. Ibid., 109 n.
9. Cf. Plutarch, *Moralia,* tr. W.W. Goodwin (Boston 1878), Bk. II, ch i, "Of the World," 132: "Pythagoras was the first philosopher that gave the name of kosmos to the world, from the order and beauty of it; for so that word signifies. Thales and his followers say that the world is one. Democritus, Epicurus, and their scholar Metrodorus affirm that there are infinite worlds

in an infinite space, for that infinite vacuum in its whole extent contains them. Empedocles, that the circle which the sun makes in its motion circumscribes the world, and that circle is the utmost bound of the world. Seleucus, that the world knows no limits. Diogenes, that the universe is infinite, but this world is finite. The Stoics make a difference between that which is called the universe, and that which is called the whole world;—the universe is the infinite space considered with the vacuum, the vacuity being removed gives the right conception of the world; so that the universe and the world are not the same thing."

10. Fletcher, 109.
11. Salisbury II, 18.
12. John points this out in Polycraticus, II, 14: "Quis nescit somniorus varias esse signifactiones, quas et usus approbat et maiorum confirmat auctoritas" [because of the varied significations of dreams, approved by experience and confirmed by the authority of our ancestors]
13. Tim Meadowcroft, "Metaphor, Narrative, Interpretation, and Reader in Daniel 2–5," in *Narrative,* ed. James Phelan, October 2000. 263
14. Daniel 2:19.
15. Daniel 2:2.
16. Daniel 2:5–6.
17. Recall that, for the dream interpreter, it is important to know details such as when the dreamer had the dream, where the dreamer slept, what food or drink was consumed prior to sleeping, and all other details that could lead to an interpretation of the dream linked to the body.
18. Krieger, *Ekphrasis,* 1. William Harmon and C. Hugh Holman's *A Handbook To Literature* provides a more detailed definition of the term: *ekphrasis* is "used in reference to the representation of an artwork of any kind in a literary work, such as a poem or string quartet inside a novel, but usually restricted to the representation of a visual or graphic work inside a literary work. The graphic work may be a painting, statue, tapestry, window, shield, urn, or other such potentially representational artifact" (Harmon 177).
19. Daniel 2:31–33.
20. Daniel 2:34.
21. Daniel 2:35.
22. Gérard Genette, *Narrative Discourse: An Essay in Method,* trans. Jane E. Lewin (Ithaca: Cornell University Press, 1980), 33.
23. Ibid., 34.
24. Ibid., 40.
25. Indeed, while we read the text, the instance of narration relies on a successful oral communication of the dream. Even as Daniel recounts the dream to Nebuchadnezzar, the king's executioner, Arioch, stands nearby.
26. Auerbach, 5–6.

27. Ibid., 6.
28. Ibid., 4.
29. Ibid.
30. Genesis 22:7.
31. Genesis 22:2.
32. Qtd. in Thorndike, *History*, 261.
33. Daniel 4:25.
34. Daniel 4:24.
35. Thorndike, *History*, 161.
36. Salisbury II, 17.
37. Ibid., 17.
38. Auerbach, 14.
39. Ibid., 14–15.
40. Ibid., 15.
41. Ibid.
42. Ibid.
43. Ibid.
44. Ibid., 15–16.
45. Qtd. in Thorndike, *History*, 163.
46. Jordan, 3.
47. Salisbury II, 17.
48. Ibid.
49. Ibid., 18.

NOTES TO CHAPTER SIX

1. Kruger, 34.
2. As J.V. Field notes in *Kepler's Geometrical Cosmology*, "Kepler shared with Plato the belief that the Universe was in some sense an expression of the nature of the God (or, more probably, in Plato's case, gods) who created it. They also shared the belief that the creator was a geometer" (Field 16).
3. Fernand Hallyn's section on the *Somnium* in his *The Poetic Structure of the World* can be claimed, quite rightly, as the work which has most greatly influenced literary analysis of the *Somnium*. Ladina Bezzola Lambert, Mary Baine Campbell, James J. Paxson, and Albert Schirrmeister have likewise contributed to the study of this minor but fascinating and important work.
4. While there has been a spate of criticism written on the *Somnium* within the past decade, the authoritative English translation of the text appeared in 1967. This translation, entitled *Kepler's Somnium: The Dream, or Posthumous Work on Lunar Astronomy*, by Edward Rosen, is the principal translation featured in this study. Another edition, featuring a translation by Patricia Frueh Kirkwood and an interpretation by John Lear, appeared in 1965.

Both of these works are currently out of print. Other English translations of the text include an unpublished translation from 1947 by Joseph Keith Lane and a partial English version by Everett F. Bleiler which appeared in the anthology *Beyond Time and Space,* an anthology edited by the noted science fiction writer and colleague of H.P. Lovecraft, August Derleth.

5. Holton, 53.
6. Anna Marie E. Roos, Luminaries in the Natural World: The Sun and the Moon in England, 1400–1720 (New York: Peter Lang, 2001), 128.
7. Donne, 3.
8. See Mary Baine Campbell's *Wonder and Science: Imagining Worlds in Early Modern Science.*
9. Alexandre Koyré, *The Astronomical Revolution: Copernicus—Kepler—Borelli,* trans. R.E.W. Maddison (Ithaca: Cornell University Press, 1973), 120.
10. Bulfinch, 29–31.
11. This attitude even filters over to literature, the last refuge of the mysterious and unknown. Consider Bertolt Brecht's *Galileo:* this text is a prime example of Galileo's deification in the face of religious orthodoxy.
12. Paul Oskar Kristeller, *Renaissance Thought and its Sources,* ed. Michael Mooney (New York: Columbia University Press, 1979), 164.
13. Ladina Bezzola Lambert, *Imagining the Unimaginable: The Poetics of Early Modern Astronomy* (Amsterdam: Rodopi, 2002), 72.
14. Ibid.
15. The essay was never formally presented, though. Rosen notes that "Veit Müller, the professor in charge of those academic exercises, was so unalterably opposed to Copernicanism that he refused to permit Kepler's theses to be heard" (Kepler, *Somnium* xvii). Kepler refers to the origins of the *Somnium* in this dissertation at Tübingen and to Müller specifically in note 2 (Kepler, *Somnium* 32).
16. See R.S. Westman's "Michael Maestlin's Adoption of the Copernican Theory" for a discussion of Maestlin's intellectual background and approach.
17. Max Caspar, *Kepler,* trans. C. Doris Hellman (New York: Dover, 1993), 47.
18. Qtd. in Caspar, 47.
19. Lambert, 68.
20. James Voelkel contests Kepler's own view that the witch trial was a direct result of the *Somnium,* however. Instead, he asserts that "Frau Kepler's trouble was rooted in her unpleasant and meddling personality" (Voelkel 99). See pp. 95–111 of Voelkel's *Johannes Kepler and the New Astronomy* for a full discussion of the proceedings of the witch trial.
21. Lawrence Lipking, "The Marginal Gloss," *Critical Inquiry* 3, no.4 (1977): 639.
22. Nicolson, *Imagination,* 64.
23. Johannes Kepler, *Kepler's Somnium,* ed. and trans. Edward Rosen (Madison: University of Wisconsin Press, 1967), 11.

24. Caspar, 149.
25. Ibid.
26. Ibid.
27. Ibid., 150.
28. R.J.W. Evans, *Rudolph II and His World* (Oxford: Clarendon Press, 1973), 196.
29. Caspar, 186.
30. Ibid., 187.
31. Kepler, *Somnium,* 11.
32. Ibid.
33. Perhaps this is the same weather system that initially prevented Kepler from seeing the new star of 1604.
34. Albert Schirrmeister, "Traum und Wissen in der Frühen Neuzeit," *Zeitsprünge* 5 (2001): 2. Translations mine unless otherwise indicated.
35. Timothy J. Reiss, *The Discourse of Modernism* (Ithaca: Cornell University Press, 1982), 143–44.
36. See my chapter "The Process of Stellification" for a discussion of the generic characteristics of the cosmological dream allegory.
37. Although perhaps an unwilling one. Caspar notes the following regarding Kepler's attitude toward horoscopes: "Casting horoscopes is for him, it is true, 'an unpleasant and at this time very begrimed work,' which one permits oneself to use only with great discretion" (Caspar 152).
38. J.V. Field, *Kepler's Geometrical Cosmology* (London: Athlone Press, 1988), xviii.
39. Schirrmeister, 3.
40. Kepler, *Somnium,* 30.
41. Ibid.
42. Giuseppe Mazzotta, *Dante, Poet of the Desert: History and Allegory in the Divine Comedy* (Princeton: Princeton University Press, 1979), 252.

NOTES TO CHAPTER SEVEN

1. The inscribed magic circle in literature serves as a dual figure: the trickster Mak of the Wakefield *Second Shepherd's Play* creates a circle around the three shepherds of the New Testament nativity narrative in order to soporize them and then steal their sheep; but Stoker's Van Helsing uses a magic circle to protect Mira Harker from the three vampire wives in *Dracula.*
2. Holton, 54.
3. Ibid., 54.
4. Marjorie Nicolson, *Voyages to the Moon* (New York: Macmillan, 1948), 41.
5. Roger Bozzetto, "Kepler, naissance de la vise spéculative fondée sur la science au sens moderne du terme," *Quarante Deux* http://www.quarantedeux.org/archives_stellaires/roger_bozzetto. Translations mine unless otherwise indicated.

6. Arthur Koestler, *The Sleepwalkers: A History of Man's Changing Vision of the Universe* (New York: Macmillan, 1959), 421.
7. A phrase used by Mary Baine Campbell, who likewise asserts the significance of the narrative form, and not merely the information it contains.
8. Although, as Karl Vossler argues in *Mediaeval Culture,* Goethe's medievalism hints at a profound connection to Dante and the view, untenable for moderns, that "existence acquire[s] its full meaning [. . .] only when beheld through the veil of vision" (Vossler 6).
9. Giorgio de Santillana, *The Age of Adventure: The Renaissance Philosophers* (New York: Mentor, 1956), 191.
10. Wolfgang von Goethe, *Goethe's Faust,* trans. Walter Kaufmann (New York: Doubleday, 1963), 2.
11. Goethe presumably includes this statement to contextualize his Faust.
12. Holton, 54.
13. Ibid., 53.
14. Ibid.
15. However, while "The *Principia* provided a standard for doing scientific investigations [. . .] Newton's influences included alchemy, unorthodox religious ideas, and a belief that God had given the ancients the true secrets of science and religion" (Tulloch 328).
16. W. Roy Laird, "Galileo and the Mixed Sciences," in *Method and Order in Renaissance Philosophy of Nature: The Aristotle Commentary Tradition,* eds. Daniel A. DiLiscia, Eckhard Kessler, and Charlotte Methuen (Aldershot: Ashgate, 1997), 253.
17. Eckhard Kessler, "Introduction," in *Method and Order in Renaissance Philosophy of Nature: The Aristotle Commentary Tradition,* eds. Daniel A. DiLiscia, Eckhard Kessler, and Charlotte Methuen (Aldershot: Ashgate, 1997), vii.
18. Bruce Stephenson, *Kepler's Physical Astronomy* (Princeton: Princeton University Press, 1994), 91.
19. Anna Marie E. Roos, *Luminaries in the Natural World: The Sun and the Moon in England, 1400–1720* (New York: Peter Lang, 2001), 31.
20. Erwin Panofsky, *Galileo as a Critic of the Arts* (The Hague: M. Nijhoff, 1954), 24–25.
21. Stephenson, 91.
22. Hallyn, *Structure,* 114.
23. Dean Swinford, "Girard Desargues and Projective Geometry," in *Science and its Times: Understanding the Social Significance of Scientific Discovery. Vol 3: 1450–1699,* ed. Neil Schlager (Detroit: Gale Group, 2000), 253.
24. Fernand Hallyn, "Du Monde de Kepler comme Anamorphoses," *Communication & Cognition* 14, 2/3 (1981), 168. Translations mine unless otherwise indicated.
25. Hallyn, *Structure,* 114.

26. Charles B. Schmitt, "Philosophy and Science in Sixteenth-Century Universities: Some Preliminary Comments," in *Studies in Renaissance Philosophy and Science* (London: Variorum Reprints), 490.
27. Richard S. Westfall, *The Construction of Modern Science: Mechanisms and Mechanics* (Cambridge: Cambridge University Press, 1977), 1.
28. Job Kozhamthadam, "The Discovery of the Laws of Kepler: A Study in the Interaction Among Empirical Science, Philosophy, and Religion" (Ph.D. diss., University of Maryland, College Park, 1988), 30.
29. Boitani, 238.
30. Ibid., 28.
31. Dante, *Paradiso* XXXIII, 133–39.
32. Curtius, 205.
33. Ibid.
34. Hallyn, *Structure,* 56.
35. Nicolaus Copernicus, *On the Revolutions of the Heavenly Spheres,* trans. A.M. Duncan (New York: Barnes & Noble, 1976), 7.
36. Hallyn, *Structure,* 55.
37. Qtd. in Hallyn, *Structure,* 56.
38. Qtd. in Kuhn, 107.
39. Kuhn, 123.
40. Ibid., 121.
41. Ibid., 121–22.
42. Hallyn, *Structure,* 57.
43. Ibid.
44. Qtd. in Caspar, 136.
45. Holton, 70.
46. Thomas Kuhn, *The Copernican Revolution: Planetary Astronomy in the Development of Western Thought* (Cambridge: Cambridge University Press, 1985), 121.
47. David Bohm and F. David Peat, *Science, Order, and Creativity* (London: Routledge, 2000), 273.
48. See Edward Rosen's Review of *Panofsky's Galileo as a Critic of the Arts* in *Isis,* XLVII (1956), 78–80.
49. Caspar, 137.
50. 1.2.122–25.
51. *Somnium,* 12.
52. Ibid., 40.
53. Ibid., 12.
54. As "dur" suggests "water" in the language of Gaul and Briton.
55. *Somnium.*, 12.
56. Ibid., 17.
57. Ibid., 15.
58. Ibid.

59. Ibid., 71.
60. Ibid., 63.
61. See Bruce Clarke, *Energy Forms: Allegory and Science in the Era of Classical Thermodynamics* (Ann Arbor: University of Michigan Press, 2001), 18 for more on the importance of energy and motion in allegory.
62. Kuhn, 118.
63. Galileo Galilei, *Dialogue Concerning the Two Chief World Systems: Ptolemaic and Copernican,* trans. Stillman Drake (Berkeley: University of California Press, 1967), 469.
64. *Somnium,* 70.
65. Ibid., 71.
66. Hallyn, "Anamorphoses," 176.
67. Qtd. in Holton, 57.
68. Caspar, 68.
69. Qtd. in Caspar, 94.
70. Caspar, 67.
71. Ibid.
72. Hallyn, "Anamorphoses," 177.
73. Roos, 69.
74. Boitani, 223.
75. Hallyn, "Anamorphoses," 177.
76. Qtd. in Hallyn, *Structure,* 161.
77. Hallyn, *Structure,* 162.
78. Bozzetto.
79. Bruce Clarke, *Allegories of Writing: The Subject of Metamorphosis* (Albany: State University of New York Press, 1995), 23.
80. Ibid.

NOTES TO CHAPTER EIGHT

1. Theresa M. Kelley, *Reinventing Allegory* (Cambridge: Cambridge University Press, 1997), 10.
2. Ernst Cassirer, *The Individual and the Cosmos in Renaissance Philosophy,* trans. Mario Domandi (Mineola: Dover Publications, 2000), 1.
3. A.G.H. Bachrach, "Luna Mendax: Some Reflections on Moon-Voyages in Early Seventeenth-Century England," in *Between Dream and Nature: Essays on Utopia and Dystopia,* ed. Dominic Baker-Smith and C.C. Barfoot (Amsterdam: Rodopi, 1987), 71.
4. Lambert, 76.
5. Mieke Bal, *Narratology: Introduction to the Theory of Narrative,* trans. Christine van Boheemen (Toronto: University of Toronto Press, 1994), 95.
6. Ibid., 76.
7. Ibid.

8. Ibid.
9. Kepler, *Somnium*, 12.
10. Ibid.
11. Ibid., 11.
12. Lambert, 76.
13. Ibid., 77.
14. Hallyn, *Structure*, 257.
15. Denise Albanese, *New Science, New World* (Durham: Duke University Press, 1996), 48.
16. Ibid., 50.
17. Lambert, 77.
18. Ibid.
19. Ibid.
20. Kepler, *Somnium*, 151.
21. Ibid., 152.
22. Lambert, 77-78.
23. Kepler, *Somnium*, 17.
24. Hallyn, *Structure*, 256.
25. Artemidorus, 28.
26. Ibid.
27. Ibid.
28. Macrobius, 92.
29. Ibid.
30. Ibid., 90.
31. Kruger, 86.
32. Caspar, 18.
33. Kruger, 87.
34. Ibid.
35. Caspar, 181.
36. Ibid., 184.
37. Cassirer, vii.
38. Caspar, 60.
39. Indeed, the completion of the Rudolphine Tables, tables of the positions of the planets which were used by astrologers and calendar makers, competed with the *Astronomia nova* for Kepler's time and attention (Caspar 140).
40. Caspar, 181.
41. Caspar, 182.
42. Ibid.
43. John Robert Christianson, *On Tycho's Island: Tycho Brahe and His Assistants, 1570-1601* (Cambridge: Cambridge University Press, 2000), 303.
44. Caspar, 185.
45. Kepler, *Somnium*, 38.

46. Ibid.
47. Hallyn, *Structure*, 258.
48. Campbell, 234.
49. Ibid., 233.
50. Ibid.
51. Lambert, 81.
52. Ibid., 82.
53. Reiss, 149.
54. Hallyn, *Structure*, 268.
55. Ibid., 12.
56. Ibid., 36.
57. Ibid., 43.
58. Ibid., 45.
59. Ibid., 13.
60. Ibid.
61. Ibid., 49.
62. Hallyn, *Structure*, 268.
63. Ibid.
64. Ibid., 269.
65. Ibid.
66. Ibid.
67. Ibid.
68. Ibid.
69. Cassirer, 121.
70. Hallyn, *Structure*, 269-70.
71. Ibid., 270.
72. David H. Levy, *Cosmic Discoveries: The Wonders of Astronomy* (Amherst: Prometheus Books, 2001), 38.
73. Lambert, 83.
74. James J. Paxson, "Kepler's Allegory of Containment, the Making of Modern Astronomy, and the Semiotics of Mathematical Thought," *Intertexts* 3, no. 2 (1999): 112.
75. Ibid., 111.
76. Ibid., 112.
77. Ibid., 117.
78. Ibid., 119.
79. Ibid., 114.
80. Ibid.
81. Hallyn, *Structure*, 279.
82. Kepler, *Somnium*, 32.
83. Lambert, 103.
84. Ibid., 105.
85. Ibid.

86. Ibid., 104.
87. Umberto Eco, *The Limits of Interpretation* (Bloomington: Indiana University Press, 1990), 13.
88. Dante Aligieri, *Literary Criticism of Dante Alighieri*, ed. and trans. Robert S. Haller (Lincoln: University of Nebraska Press, 1973), 99.
89. Hallyn, *Structure*, 264.
90. Ibid., 265.
91. Ibid., 280.
92. Ibid.
93. Lambert, 98.

NOTES TO CHAPTER NINE

1. Stillman Drake, *Telescopes, Tides, and Tactics: A Galilean Dialogue About the Starry Messenger and Systems of the World* (Chicago: University of Chicago Press, 1983), 61.
2. William Shea, "Looking at thc Moon as Another Earth. Terrestrial Analogies and Seventeenth-Century Telescopes," in *Metaphor and Analogy in the Sciences*, ed. Fernand Hallyn (Dordrecht: Kluwer Academic Publishers, 2000), 96.
3. Field, 35.
4. Qtd. in Shea, 98-99.
5. Harriot was directly involved in attempts to gain a greater understanding of the New World. His *A Briefe and True Report of the New Found Land of Virginia*, first published in 1588, provides an account of the Algonkian tribe of Virginia. His descriptions, besides encouraging British settlement, seek to position the natives within a Christian historical context. As Andrew Hadfield notes in "Thomas Harriot and John White: Ethnography and Ideology in the New World," "No identity [for these previously unknown peoples] could be established without recourse to a theological explanation" (Hadfield 201).
6. Roos, 93-94.
7. Qtd. in Ewen A. Whitaker, "Selenography in the Seventeenth Century," in *Planetary Astronomy from the Renaissance to the Rise of Astrophysics. Part A: Tycho Brahe to Newton*, eds. R. Taton and C. Wilson (Cambridge: Cambridge University Press, 1989), 120.
8. Shea, 98.
9. Ibid.
10. We cannot, however, undermine the significance of mathematical achievement. As Stephen Jay Gould reminds us in "Happy Thoughts on a Sunny Day in New York City," "Galileo described the universe in his most famous line: 'This grand book is written in the language of mathematics, and its characters are triangles, circles, and other geometrical figures'" (Gould 3).

Still, as Gould admonishes a scientific establishment "oversold on the mathematical precision of nature," he contends that "much of nature is messy and multifarious, markedly resistant to simple mathematical expression" (Gould 4,3).

11. Clarke, *Energy*, 28.
12. Paul De Man, "'Conclusions': Walter Benjamin's 'The Task of the Translator,'" in *The Resistance to Theory* (Minneapolis: University of Minnesota Press, 1986), 87.
13. David L. Clark, "Monstrosity, Illegibility, Denegation: De Man, bp Nichol, and the Resistance to Postmodernism," in *Monster Theory*, ed. Jeffrey Jerome Cohen (Minneapolis: University of Minnesota Press, 1996), 41.
14. Kepler, *Somnium*, 17.
15. Mary Baine Campbell, "Alternative Planet: Kepler's Somnium (1634) and the New World," in *The Arts of 17th Century Science: Representations of the Natural World in European and North American Culture*, eds. Claire Jowitt and Diane Watt (Aldershot: Ashgate, 2002), 236.
16. Ibid.
17. Ibid., 237.
18. Campbell, 237.
19. Though, as Kepler postulated, the moon was capable of influencing the tides.
20. Jeffrey Jerome Cohen, "Monster Culture (Seven Theses)," in *Monster Theory*, ed. Jeffrey Jerome Cohen (Minneapolis: University of Minnesota Press, 1996), 4.
21. Kepler, *Somnium*, 50.
22. Ibid.
23. Ibid.
24. Ibid.
25. Ibid.
26. Joel Fineman, "The Structure of Allegorical Desire," in *Allegory and Representation*, ed. Stephen J. Greenblatt (Baltimore: Johns Hopkins University Press, 1981), 8.
27. Kepler, *Somnium*, 39.
28. Ibid., 50.
29. P. Lévy, *Les Technologies de l'intelligence* (Paris: La Découverte, 1990), 209. Translations mine.
30. Lambert, 91.
31. Kepler, *Somnium*, 14.
32. John Block Friedman, *The Monstrous Races in Medieval Art and Thought* (Cambridge: Harvard University Press, 1981), 32.
33. Kepler, *Somnium*, 60.
34. Ibid.
35. Ibid.

36. Ibid.
37. Ibid.
38. Ibid., 62.
39. Roland Barthes, *Image Music Text*, trans. Stephen Heath (New York: Hill and Wang, 1977), 181.
40. Ibid.
41. Christian Bök, *'Pataphysics: The Poetics of an Imaginary Science* (Evanston: Northwestern University Press, 2002), 3.
42. Barthes, 181-82.
43. Barthes, 182.
44. Kepler, *Somnium*, 57.
45. Ibid.
46. Giovanni Battista della Porta popularized the camera obscura with his *Magiae Naturalis* (1558). The initial demonstration to his invention indicates that Kepler refers to the supernatural just by referring to the camera obscura. After perfecting his invention, "della Porta summoned his friends and important members of Naples society to his home for a demonstration. Instead of sharing his excitement, the group was appalled when they saw real human images displayed on the wall, believing it to be the work of witchcraft. The Catholic Church got wind of della Porta's demonstration and promptly charged him with sorcery. His work was banned for six years." For a clear and concise overview of the camera obscura, see Stephanie Watson, "Camera Obscura: Ancestor of Modern Photography," in *Science and its Times: Understanding the Social Significance of Scientific Discovery. Vol. 3: 1450-1699*, ed. Neil Schlager (Detroit: Gale Group, 2000), 423-426.
47. Kepler, *Somnium*, 57.
48. Ibid.
49. Ibid., 53.
50. Ibid.
51. In this section of the Book of Daniel, Daniel derives his authority, given by God and affirmed by man, from his textual interpretation skills. After he decodes the inscription, we read in Daniel 6:29 that "Daniel was clothed in purple, a gold chain was placed around his neck, and he was proclaimed the third highest ruler in the kingdom."
52. Kepler, *Somnium*, 57.
53. Whitman, *Allegory*, 72.
54. Ibid.
55. Ibid., 71.
56. Ibid., 72.
57. Ibid.
58. Ibid.
59. Ibid.

60. Ibid.
61. Ibid., 73.
62. Apulius' viewpoint on this matter closely parallels that of Origen. Thus, "When Apuleius speaks of the demons in general terms, he does so by contrasting their nature both with higher principles: the gods and with lower ones: human souls. Thus, he repeatedly stresses the notion that they are 'intermediate powers' (*mediae potestates, medioximi*)" (Gersh 309).
63. Whitman, *Allegory*, 73.
64. Gersh notes that, for a Platonist such as Apuleius, there is a group of demons "which are never incarnate in human bodies. As in the case of the gods, we find these spiritual beings described from both a subjective and an objective viewpoint, the former finding expression in Apuleius' assertion that they are 'visible to nobody' (nemini conspicui)—an account negative in character—or that they can reveal themselves as 'a kind of voice, although not the usual or human kind' (*vox quaepiam . . . non usitata vox nec humana*)" (Gersh 312). These characteristics found in Apuleius' writings "can be extracted from Plato's dialogues" (Gersh 312).
65. Whitman, *Allegory*, 79.
66. Ibid.
67. Ibid., 80.
68. Curtius, 319.
69. Qtd. in Curtius, 319.
70. James J. Paxson, "Revisiting the Deconstruction of Narratology: Master Tropes of Narrative Embedding and Symmetry," *Style* 35, no.1 (2001): 141.
71. Kepler, *Somnium*, 117.
72. Qtd. in Caspar, 127.
73. Caspar, 130.
74. Paxson, "Deconstruction," 142.
75. Bal, 141.
76. Anthony Grafton, *The Footnote: A Curious History* (London: Faber and Faber, 1997), 5.
77. Ibid., 7.
78. Jacques Derrida, "This is Not An Oral Footnote," in *Annotation and its Texts*, trans. and ed. Stephen A. Barney (Oxford: Oxford University Press, 1991), 196.
79. Lambert draws a different conclusion that does not take this asymmetry into account. Instead, she argues that:

> The complex correspondences of characters and settings that echo through the entire narrative of the Dream cause the differentiation of both narrative levels and places to collapse: the parallel arrangement of literally all the characters that figure

> in the narrative (be it with regard to their function in the plot, their character traits or what they experience) combined with the similarities between the various settings (Iceland, Hven, the Moon) seem to imply that the separate parts of the narrative are somehow the same, that they are mere variants of each other. (Lambert 87)

80. Paxson, "Deconstruction," 142.
81. Kepler, *Somnium*, 28-29.
82. Ibid., 11.
83. Wetherbee, 220.
84. Ibid., 221.
85. Ibid.
86. Ibid.
87. Ibid.
88. Ibid.
89. See Southern, 61-85.
90. William of Conches' view on the world soul transformed through his career: "Scholars have been fascinated by William's changing attitude towards the world-soul described by Plato in the *Timaeus*. His works reveal an early confidence in identifying it with the Holy Spirit, followed by greater caution, until by the time of the *Dragmaticon* the world-soul receives not so much as a mention" (Elford 326).
91. Indeed, D.E. Luscombe notes that, for Abelard, "when properly interpreted, Plato—*maximus philosophorum*—and his followers may be seen to have expressed the mystery of the Trinity (*totius Trinitatis summam post prophetas patenter ediderunt*)" (Luscombe 302).
92. Thierry "finds the moving power inherent in the universe understood as spirit by the pagan philosophers, Hermes (whose testament, *Asclepius*, was deemed to be a work of immense antiquity), Plato, and Virgil (the inspired *vates* who speaks of 'the spirit within'), just as it is by the biblical prophets, Moses, David, and Solomon. And it is this same power, Thierry concludes triumphantly, that Christians call 'the Holy Spirit'" (Dronke, "Thierry" 379). Still, in some senses, Thierry anticipates the work of scientists like Kepler. He affirms that "the cosmos is no mere mirror-image of divinity" and instead "works 'in accordance with physics (*secundum physicam*)'"(383).
93. Hallyn, *Structure*, 278.
94. Kepler, *Somnium*, 15.
95. Ibid., 17.
96. Whitman, *Allegory*, 203.
97. See my discussion of the world soul in "Language and Its Limits as a Celestial Vehicle" for a more detailed exploration of language and the divine.
98. Whitman, *Allegory*, 203.
99. Holton, 67.

100. Ibid.
101. Kepler, *Somnium*, 53.
102. While Levania exists in opposition to the sun, it is, above all, "an inverted image of Earth, but an Earth conceived according to the obsolescent Ptolemaic model of geocentrism" (Paxson, "Deconstruction" 139).
103. Holton, 69.
104. Ibid., 70.
105. Ibid.
106. Charlotte Methuen, *Kepler's Tübingen: Stimulus to a Theological Mathematics* (Aldershot: Ashgate, 1998), 162.
107. Ibid., 165.
108. Kepler studied here from 1589 until 1594. For exact dates, see Hermelink, Heinrich (ed.), *Die Matrikeon der Universität Tübingen*, vol. 1: 1477-1600. Kohlhammer, Stuttgart, 1906.
109. Methuen, 165.
110. Ibid., 164.
111. Ibid., 171.
112. Ibid.
113. Ibid., 172.
114. Ibid., 173.
115. Qtd. in Metheun, 206.
116. Metheun, 224.
117. Kepler, *Somnium*, 19,22.
118. Job 1:7.
119. Serres, 201.
120. Ibid., 223.

References

Aers, David. "Medievalists and Deconstruction: An Exemplum." *From Medieval to Medievalism*. Edited by John Simons. London: Macmillan, 1992.

Agrippa von Nettesheim, Henricus Cornelius. *Three Books of Occult Philosophy*. Edited by Willis F. Whitehead. New York, 1897.

Alain de Lille. *Anticlaudianus, or the Good and Perfect Man*. Translated by James J. Sheridan. Toronto: Pontifical Institute of Mediaeval Studies, 1973.

———. *The Plaint of Nature*. Translated and edited by James J. Sheridan. Toronto: Pontifical Institute of Mediaeval Studies, 1980.

Albanese, Denise. *New Science, New World*. Durham: Duke University Press, 1996.

Albertus Magnus. *Opera Omnia*, 38 vols. Edited by August Borgnet. Paris: Apud Ludovicum Vivès, 1890-99.

Alexandrakis, Aphrodite. "The Notion of Beauty in the Structure of the Universe: Pythagorean Influences on Plotinus." In *Neoplatonism and Nature: Studies in Plotinus' Enneads*, edited by Michael F. Wagner. Albany: SUNY Press, 2002.

Anderson, George K. *Old and Middle English Literature from the Beginnings to 1485*. New York: Collier Books, 1962.

Anonymous. *The Cloud of Unknowing*. Translated by Clifton Wolters. New York: Penguin, 1974.

Artemidorus Daldianus. *The Interpretation of Dreams: Oneirocritica*. 2nd ed. Translated and edited by Robert J. White. Torrance: Original Books, 1990.

Auerbach, Erich. *Mimesis: The Representation of Reality in Western Literature*. Translated by Willard R. Trask. Princeton: Princeton University Press, 1953.

Augustine. *On Christian Doctrine*. Translated by D.W. Robertson. Indianapolis: Bobbs-Merrill, 1958.

Bachrach, A.G.H. "Luna Mendax: Some Reflections on Moon-Voyages in Early Seventeenth-Century England." In *Between Dream and Nature: Essays on Utopia and Dystopia*, edited by Dominic Baker-Smith and C.C. Barfoot. Amsterdam: Rodopi, 1987.

Bakhtin, Mikhail. *The Dialogic Imagination*. Translated by Caryl Emerson and Michael Holquist. Austin: University of Texas Press, 1981.

Bal, Mieke. *Narratology: Introduction to the Theory of Narrative*. Translated by Christine van Boheemen. Toronto: University of Toronto Press, 1994.

Bargeliotes, Leonidas C. "Divinized and De-divinized: Conceptions of the World and of *Cosmos*." In *Neoplatonism and Western Aesthetics*, edited by Aphrodite Alexandrakis, 229-246. Albany: SUNY Press, 2001.

Barthes, Roland. *Image Music Text*. Translated by Stephen Heath. New York: Hill and Wang, 1977.

Bernardo, Aldo S. *Petrarch, Scipio and the "Africa": The Birth of Humanism's Dream*. Baltimore: Johns Hopkins Press, 1962.

Bernardus Silvestris. *The Cosmographia of Bernardus Silvestris*. Translated and Edited by Winthrop Wetherbee. New York: Columbia University Press, 1973.

Berriot, François. *Exposicions et significacions des songes et Les songes Daniel*. Genève: Librairie Droz, 1989.

———. *Spiritualités, Hétérodoxies et Imaginaires: Études sur le Moyen-Age et la Renaissance*. Saint-Étienne: Publications de l'Université de Saint-Étienne, 1994.

Bloom, Harold. "Poetry, Revisionism, Repression." In *Critical Theory Since 1965*, edited by Hazard Adams and Leroy Searle, 330-343. Tallahassee: Florida State University Press, 1986.

Boethius. *The Consolation of Philosophy*. Translated by Richard Green. New York: Macmillan, 1962.

Bohm, David and F. David Peat. *Science, Order, and Creativity*. 2nd ed. London: Routledge, 2000.

Boitani, Piero. *The Tragic and the Sublime in Medieval Literature*. Cambridge: Cambridge University Press, 1989.

Bök, Christian. *'Pataphysics: The Poetics of an Imaginary Science*. Evanston: Northwestern University Press, 2002.

Bollack, Mayotte. *La raison de Lucrèce: Constitution d'une poétique philosophique avec un essai d'interprétation de la critique lucrétienne*. Paris: Les Éditions de Minuit, 1978.

Bono, J. J. *The Word of God and the Languages of Man: Interpreting Nature in Early Modern Science and Medicine*. Madison: University of Wisconsin Press, 1995.

Borges, Jorge Luis. *Libro de Sueños*. Madrid: Alianza Editorial, 2000.

Bozzetto, Roger. "Kepler, naissance de la visée spéculative fondée sur la science au sens moderne du terme." http://www.quarente-deux.org/archives_stellaires/roger_bozzetto/

Brittan, Simon. *Poetry, Symbol, and Allegory: Interpreting Metaphorical Language from Plato to the Present*. Charlottesville: University of Virginia Press, 2003.

Brod, Max. *Tycho Brahes Weg Zu Gott*. Leipzig: Kurt Wolff Verlag, 1915.

Brown, P. *Reading Dreams: The Interpretation of Dreams from Chaucer to Shakespeare*. Oxford: Oxford University Press, 1999.

Brunel, Pierre. *Histoire de la Littérature Française du Moyen Âge au XVIII siècle*. Paris: Bordas, 2001.

Bulfinch, Thomas. *Bulfinch's Mythology*. New York: Gramercy Books, 1979.

Burnett, Charles. "Scientific Speculations." In *A History of Twelfth-Century Western Philosophy*, edited by Peter Dronke, 151-176. Cambridge: Cambridge University Press, 1988.

Burrow, J.A. *Medieval Writers and Their Work: Middle English Literature and its Background 1100-1500*. Oxford: Oxford University Press, 1982.

Calvino, Italo. *The Castle of Crossed Destinies*. Translated by William Weaver. NewYork: Harvest, 1979.

Campbell, Mary Baine. "Alternative Planet: Kepler's Somnium (1634) and the New World." In *The Arts of 17th Century Science: Representations of the Natural World in European and North American Culture*, edited by Claire Jowitt and Diane Watt, 232-249. Aldershot: Ashgate, 2002.

Caspar, Max. *Kepler*. Translated by C. Doris Hellman. New York: Dover, 1993.

Cassirer, Ernst. *The Individual and the Cosmos in Renaissance Philosophy*. Translated by Mario Domandi. Mineola: Dover Publications, 2000.

Carilla, Emilio. *Jorge Luis Borges Autor de "Pierre Ménard" (Y Otros Estudios Borgesianos)*. Bogotá: Biblioteca de Publicaciones del Instituto Caro y Cuervo, 1989.

Cazclles, Henri. *Introduction à la Bible: Introduction Critique à l'Ancien Testament*. 2 vols. Paris: Desclée et Cie, 1973.

Charity, A.C. *Events and their Afterlife: The Dialectics of Christian Typology in the Bible and Dante*. Cambridge: Cambridge University Press, 1966

Chaucer, Geoffrey. *The Riverside Chaucer*. Edited by Larry D. Benson. Boston: Houghton Mifflin Company, 1987.

Christianson, John Robert. *On Tycho's Island: Tycho Brahe and His Assistants, 1570-1601*. Cambridge: Cambridge University Press, 2000.

Clark, David L. "Monstrosity, Illegibility, Denegation: De Man, bp Nichol, and the Resistance to Postmodernism." In *Monster Theory*, edited by Jeffrey Jerome Cohen, 40-71. Minneapolis: University of Minnesota Press, 1996.

Clarke, Bruce. *Allegories of Writing: The Subject of Metamorphosis*. Albany: State University of New York Press, 1995.

———. *Energy Forms: Allegory and Science in the Era of Classical Thermodynamics*. Ann Arbor: University of Michigan Press, 2001.

Clifford, Gay. *The Transformations of Allegory*. London: Routledge, 1974.

Cogan, Marc. *The Design in the Wax: The Structure of the Divine Comedy and Its Meaning*. Notre Dame: University of Notre Dame Press, 1999.

Cohen, Jeffrey Jerome. "Monster Culture (Seven Theses)." In *Monster Theory*, edited by Jeffrey Jerome Cohen, 3-25. Minneapolis: University of Minnesota Press, 1996.

———. "Preface: In a Time of Monsters." In *Monster Theory*, edited by Jeffrey Jerome Cohen, vii-xiii. Minneapolis: University of Minnesota Press, 1996.

———. "The Order of Monsters: Monster Lore and Medieval Narrative Traditions." In *Telling Tales: Medieval Narratives and the Folk Tradition*, edited by Francesca Canadé Sautman, Diana Conchado, and Giuseppe Carlo Di Scipio, 37-58. New York: St. Martin's Press, 1998.

Collins, Anne. "Art, Technology, and the American Space Program, 1962-1972." *Intertexts* 3:2 (Fall 1999): 124-146.

Connor, James A. *Kepler's Witch: An Astronomer's Discovery of Cosmic Order Amid Religious War, Political Intrigue, and the Heresy Trial of His Mother*. New York: HarperCollins, 2004.

Copeland, Rita. *Rhetoric, Hermeneutics, and Translation in the Middle Ages: Academic Traditions and Vernacular Texts*. Cambridge, Cambridge University Press, 1991.

Crombie, A.C. *Medieval and Early Modern Science*. Vol 1. New York: Doubleday, 1959.

Crum, Ralph B. *Scientific Thought in Poetry*. New York: AMS Press, Inc., 1966.

Cumont, Franz. *Astrology and Religion Among the Greeks and Romans*. New York: 1912.

Curtius, Ernst Robert. *European Literature and the Latin Middle Ages*. Translated by Willard R. Trask. Princeton: Princeton University Press, 1953.

Dante Alighieri. *Literary Criticism of Dante Alighieri*. Translated and edited by Robert S. Haller. Lincoln: University of Nebraska Press, 1973.

———. *Paradiso*. Translated by Allen Mandelbaum. New York: Bantam Books, 1982.

De Lubac, Henri. *Medieval Exegesis. Vol. 1. The Four Senses of Scripture*. Translated by Mark Sebanc. Grand Rapids: William B. Eerdmans, 1998.

De Man, Paul. *The Resistance to Theory*. Minneapolis: University of Minneapolis Press, 1986.

Demonet, Marie-Luce. *Les Voix du Signe: Nature et Origine du Langage à la Renaissance (1480-1580)*. Paris: Librairie Honoré Champion, 1992.

Derrida, Jacques. "The Law of Genre." Translated by Avital Ronell. In *On Narrative*, edited by W.J.T. Mitchell, 51-77. Chicago: University of Chicago Press, 1981.

———. "Dialanguages." Translated by Peggy Kamuf. In *Points*, edited by Elisabeth Weber, 132-155. Stanford: Stanford University Press, 1995.

———. *Glas*. Translated by John P. Leavey, Jr., and Richard Rand. Lincoln: University of Nebraska Press, 1986.

———. *Of Grammatology*. Translated by Gayatri C. Spivak. Baltimore: Johns Hopkins University Press, 1976.

———. *Signéponge*. Paris: Éditions de Seuil, 1988.

Drake, Stillman. *Telescopes, Tides, and Tactics: A Galilean Dialogue About the Starry Messenger and Systems of the World*. Chicago: University of Chicago Press, 1983.

Dronke, Peter. *Fabula: Explorations Into the Uses of Myth in Medieval Platonism*. Brill: Leiden, 1974.

———. *Sources of Inspiration: Studies in Literary Transformations, 400-1500*. Rome: Edizioni di Storia e Letteratura, 1997.

———. "Thierry of Chartres." In *A History of Twelfth-Century Western Philosophy*, edited by Peter Dronke, 358-385. Cambridge: Cambridge University Press, 1988.

Dubost, Francis. *Aspects Fantastiques de la Littérature Narrative Médiévale (XIIème-XIIIème siècles): L'Autre, l'Ailleurs, l'Autrefois*. 2 vols. Paris: Librairie-Honoré Champion, 1991.

Dulaey, Martine. *Le Rêve dans la vie et la pensée de Saint Augustin*. Paris: Études Augustiniennes, 1973.

Eco, Umberto. *The Island of the Day Before*. Translated by William Weaver. New York: Penguin Books, 1995.

Edwards, Robert R. *The Dream of Chaucer: Representation and Reflection in the Early Narratives*. Durham: Duke University Press, 1989.

Elford, Dorothy. "William of Conches." In *A History of Twelfth-Century Western Philosophy*, edited by Peter Dronke, 308-327. Cambridge: Cambridge University Press, 1988.

Evans, Murray J. "Piers Plowman and the Sublime." *Exemplaria* 9.2 (1997): 421-440.

Evans, R.J.W. *Rudolph II and His World*. Oxford: Clarendon Press, 1973.

Fagan, Brian. *The Little Ice Age: How Climate Made History, 1300-1850*. New York: Basic Books, 2002.

Fairlamb, Horace. *Critical Conditions*. Cambridge: Cambridge University Press, 1994.

Ferguson, Kitty. *Tycho and Kepler: The Unlikely Partnership That Forever Changed Our Understanding of the Heavens*. New York: Walker and Company, 2002.

Field, J.V. *Kepler's Geometrical Cosmology*. London: Athlone Press, 1988.

Fineman, Joel. "The Structure of Allegorical Desire." In *Allegory and Representation*, edited by Stephen J. Greenblatt, 26-60. Baltimore: Johns Hopkins University Press, 1981.

Fletcher, Angus. *Allegory: The Theory of a Symbolic Mode*. Ithaca: Cornell University Press, 1964.

Fowler, Alastair. "The Future of Genre Theory: Functions and Constructional Types." In *The Future of Literary Theory*, edited by Ralph Cohen. New York: Routledge, 1989.

Friedman, John Block. *The Monstrous Races in Medieval Art and Thought*. Cambridge: Harvard University Press, 1981.

Gadamer, Hans-Georg. *Truth and Method*. 2nd ed. Translated by Joel Weinsheimer and Donald G. Marshall. New York: Crossroad Publishing Corporation, 1989.

Galileo Galilei. *Sidereus Nuncius or, The Sidereal Messenger*. Translated by Albert van Helden. Chicago: University of Chicago Press, 1989.

Genette, Gérard. *The Architext: An Introduciton*. Translated by Jane E. Lewin. Berkeley: University of California Press, 1992.

———. *Narrative Discourse: An Essay in Method*. Translated by Jane E. Lewin. Ithaca: Cornell University Press, 1980.

Gersh, Stephen. *Middle Platonism and Neoplatonism: The Latin Tradition*. Vol. 1. Notre Dame: University of Notre Dame Press, 1986.

———. *Middle Platonism and Neoplatonism: The Latin Tradition.* Vol. 2. Notre Dame: University of Notre Dame Press, 1986.

Gingerich, Owen. "The Tycho Illusion: Performing the Cutout Correctly." *Psycoloquy* 9:52 (1998). http://psycprints.ecs.soton.ac.uk/archive/00000601/

Goethe, Wolfgang von. *Goethe's Faust.* Translated by Walter Kaufmann. New York: Doubleday, 1963.

Goldie, Matthew Boyd. *Middle English Literature: A Historical Sourcebook.* Oxford: Blackwell, 2003.

Goldschmidt, E.P. *Medieval Texts and Their First Appearance in Print.* Oxford: University of Oxford Press, 1943.

Goodman, Nelson. *Ways of Worldmaking.* Indianapolis: Hackett Publishing Company, 1978.

Gould, Stephen Jay. "Happy Thoughts on a Sunny Day in New York City." In *Dinosaur in a Haystack: Reflections in Natural History*, edited by Stephen Jay Gould, 3-10. New York: Harmony Books, 1995.

Grafton, Anthony. *The Footnote: A Curious History.* London: Faber and Faber, 1997.

Gregory, Tullio. "The Platonic Inheritance." In *A History of Twelfth-Century Western Philosophy*, edited by Peter Dronke, 54-80. Cambridge: Cambridge University Press, 1988.

Hadfield, Andrew. "Thomas Harriot and John White: Ethnography and Ideology in the New World." In *The Arts of 17th Century Science: Representations of the Natural World in European and North American Culture*, edited by Claire Jowitt and Diane Watt, 199-216. Aldershot: Ashgate, 2002.

Hallyn, Fernand. "Du Monde de Kepler comme Anamorphose." *Communication & Cognition.* Vol. 14, No. 2/3 (1981). Universiteit Gent. 165-198.

———. *The Poetic Structure of the World: Copernicus and Kepler.* Translated by Donald M. Leslie. New York: Zone Books, 1993.

Harmon, William and C. Hugh Holman. *A Handbook To Literature.* 8th ed. New Jersey: Prentice Hall, 2000.

Haskins, Charles Homer. *The Renaissance of the 12th Century.* New York: Meridian, 1966.

Hayles, N. Katherine. *Chaos Bound: Orderly Disorder in Contemporary Literature and Science.* Ithaca: Cornell University Press, 1990.

Heer, Friedrich. *The Medieval World: Europe 1100-1350.* Translated by Janet Sondheimer. New York: Mentor, 1962.

Hendrix, John. "Neoplatonism in the Design of Baroque Architecture." In *Neoplatonism and Western Aesthetics*, edited by Aphrodite Alexandrakis, 113-130. Albany: SUNY Press, 2001.

Hollander, Robert. *Allegory in Dante's Commedia.* Princeton: Princeton University Press, 1969.

Holquist, Michael. *Bakhtin and His World.* Austin: University of Texas Press

Holmes, George. *The Later Middle Ages, 1272-1485.* New York: Norton, 1962.

Holton, Gerald. *Thematic Origins of Scientific Thought: Kepler to Einstein*. Cambridge: Harvard University Press, 1988.

Homer. *The Odyssey*. Translated by W.H.D. Rouse. New York: Mentor, 1937.

Honnefelder, Ludger. "The Concept of Nature in Medieval Metaphysics." In *Nature in Medieval Thought: Some Approaches East and West*, edited by Chumaru Koyama, 75-93. Brill: Leiden, 2000.

Hopper, Vincent Foster. *Medieval Number Symbolism: Its Sources, Meaning, and Influence on Thought and Expression*. New York: Dover, 2000

Huizinga, Johan. *The Waning of the Middle Ages*. New York: Anchor, 1954.

Jager, Eric. *The Tempter's Voice: Language and the Fall in Medieval Literature*. Ithaca: Cornell University Press, 1993.

Jameson, Frederic. *Postmodernism, Or, the Cultural Logic of Late Capitalism*. Durham: Duke University Press, 1990.

———. *The Prison-House of Language: A Critical Account of Structuralism and Russian Formalism*. Princeton: Princeton University Press, 1972.

Jauss, Hans Robert. *Question and Answer: Forms of Dialogic Understanding*. Edited and translated by Michael Hays. Minneapolis: University of Minnesota Press, 1989.

[John of Salisbury]. *Ioannis Saresberiensis episcopi carnotensis Policratici*. 2 vols. Edited by Clemens C.I. Webb. Frankfurt: Minerva G.M.B.H., 1909.

Jordan, Robert M. *Chaucer and the Shape of Creation: The Aesthetic Possibilities of norganic Structure*. Cambridge: Harvard University Press, 1967.

Joyce, James. *A Portrait of the Artist as a Young Man*. Edited by R.B. Kershner. New York: Bedford Books, 1993.

Kelley, Theresa M. *Reinventing Allegory*. Cambridge: Cambridge University Press, 1997.

Kelly, Louis G. "Saint Augustine and Saussurean Linguistics." *Augustinian Studies* 6 (1975): 45-64.

Kent, Thomas. *Interpretation and Genre: The Role of Generic Perception in the Study of Narrative Texts*. Cranbury, NJ: Associated University Presses, 1986.

Kepler, Johannes. *Gesammelte Werke*. 13 vols. Edited by Walther von Dyck, Max Caspar et al. Munich: C.H. Beck, 1938-.

———. *Kepler's Somnium*. Translated and edited by Edward Rosen. Madison: University of Wisconsin Press, 1967.

Kessler, Eckhard. "Introduction." In *Method and Order in Renaissance Philosophy of Nature: The Aristotle Commentary Tradition*, edited by Daniel A. DiLiscia, Eckhard Kessler, and Charlotte Methuen, vii-xi. Aldershot: Ashgate, 1997.

Kesten, Hermann. *Copernicus und seine Welt*. Frankfurt: Ullstein Taschenbuch, 1983.

Klibansky, Raymond. *The Continuity of the Platonic Tradition During the Middle Ages*. London: The Warburg Institute, 1939.

Kluxen, Wolfgang. "Nature in the Ethics of the Middle Ages." In *Nature in Medieval Thought: Some Approaches East and West*, edited by Chumaru Koyama, 23-42. Brill: Leiden, 2000.

Knowles, David. *The Evolution of Medieval Thought*. New York: Vintage Books, 1962.

Koestler, Arthur. *The Sleepwalkers: A History of Man's Changing Vision of the Universe*. New York: Macmillan, 1959.

———. *The Watershed: A Biography of Johannes Kepler*. New York: Macmillan, 1960.

Koyré, Alexandre. *The Astronomical Revolution: Copernicus—Kepler—Borelli*. Translated by R.E.W. Maddison. Ithaca: Cornell University Press, 1973.

———. *Galileo Studies*. Translated by John Mepham. Atlantic Highlands: Humanities Press, 1978.

Kozhamthadam, Job. *The Discovery of the Laws of Kepler: A Study in the Interaction Among Empirical Science, Philosophy, and Religion*. Unpublished dissertation. University of Maryland College Park, 1988.

Krieger, Murray. "'A Waking Dream': The Symbolic Alternative to Allegory." In *Allegory, Myth, and Symbol*, edited by Morton W. Bloomfield. Cambridge: Harvard University Press, 1981.

Krieger, Murray. *Ekphrasis: The Illusion of the Natural Sign*. Baltimore: The Johns Hopkins University Press, 1992.

Kristeller, Paul Oskar. *Renaissance Thought and its Sources*. Edited by Michael Mooney. New York: Columbia University Press, 1979.

Kruger, Steven F. *Dreaming in the Middle Ages*. Cambridge: Cambridge University Press, 1992.

Laird, W. Roy. "Galileo and the Mixed Sciences." In *Method and Order in Renaissance Philosophy of Nature: The Aristotle Commentary Tradition*, edited by Daniel A. DiLiscia, Eckhard Kessler, and Charlotte Methuen. Aldershot: Ashgate, 1997.

Lambert, Ladina Bezzola. *Imagining the Unimaginable: The Poetics of Early Modern Astronomy*. Amsterdam: Rodopi: 2002.

Lear, J. *Kepler's Dream*. Translated by P.F. Kirkwood. Berkeley: University of California Press, 1965.

Levine, Robert. "Gower as Gerontion: Oneiric Autobiography in the *Confessio Amantis*." *Medieaevistik* 5 (1992): 81-86.

Levy, David H. *Cosmic Discoveries: The Wonders of Astronomy*. Amherst: Prometheus Books, 2001.

Lévy, P. *Les Technologies de l'intelligence*. Paris: La Découverte, 1990.

Lewis, C.S. *The Discarded Image*. Cambridge: Cambridge University Press, 1964.

Lipking, Lawrence. "The Marginal Gloss." *Critical Inquiry* 3, no. 4 (1977): 609-656.

Luscombe, D.E. "Peter Abelard." In *A History of Twelfth-Century Western Philosophy*, edited by Peter Dronke, 279-307. Cambridge: Cambridge University Press, 1988.

Lutwack, Leonard. *Birds in Literature*. Gainesville: University Press of Florida, 1994.

Lynch, Kathryn L. "Partitioned Fictions: The Meaning and Importance of Walls in Chaucer's Poetry." In *Art and Context in Late Medieval English Narrative*, edited by Robert R. Edwards, 107-126. Cambridge: D.S. Brewer, 1994.

Machamer, Peter. "The Nature of Metaphor and Scientific Description." In *Metaphor and Analogy in the Sciences*, edited by Fernand Hallyn, 35-52. Dordrecht: Kluwer Academic Publishers, 2000.

Macrobius. *Commentary on the Dream of Scipio by Macrobius*. Translated and edited by illiam Harris Stahl. New York: Columbia University Press, 1990. *The Travels of Sir John Mandeville*. Translated by C.W.R.D. Moseley. London: Penguin, 1983.

Margolis, Howard. "Tycho's Illusion: How It Lasted 400 Years, and What That Implies About Human Cognition." *Psycoloquy* 9, no. 32 (1998). http://psycprints.ecs.soton.ac.uk/archive/00000581/

Mazzeo, Joseph Anthony. *Medieval Cultural Tradition in Dante's Commedia*. Ithaca: Cornell University Press, 1960.

———. *Structure and Thought in the Paradiso*. Ithaca: Cornell University Press, 1958.

Mazzotta, Giuseppe. *Dante, Poet of the Desert: History and Allegory in the Divine Comedy*. Princeton: Princeton University Press, 1979.

Meadowcroft, Tim. "Metaphor, Narrative, Interpretation, and Reader in Daniel 2-5." *Narrative* 8 (October 2000): 257-278.

Melberg, Arne. *Theories of Mimesis*. Cambridge: Cambridge University Press, 1995.

Melchior-Bonnet, Sabine. *The Mirror: A History*. Translated by Katharine H. Jewett. New York: Routledge, 2001.

Methuen, Charlotte. *Kepler's Tübingen: Stimulus to a Theological Mathematics*. Aldershot: Ashgate Publishing Limited, 1998.

Miller, J. Hillis. "The Two Allegories." In *Allegory, Myth, and Symbol*, edited by Morton W. Bloomfield. Cambridge: Harvard University Press, 1981.

Moore, Patrick. *The Great Astronomical Revolution: 1534-1687 and the Space Age Epilogue*. Chichester: Albion, 1994.

Moran, Joe. *Interdisciplinarity*. London: Routledge, 2002.

Moretti, Franco. *Modern Epic: The World System from Goethe to García Márquez*. Translated by Quintin Hoare. London: Verso, 1996.

Newman, Francis Xavier. *Somnium: Medieval Theories of Dreaming and the Form of Vision Poetry*. Princeton: Princeton University, 1963. Unpublished diss.

Nicolson, Marjorie Hope. *Science and Imagination*. Hamden: Archon Books, 1976.

———. *Voyages to the Moon*. New York: Macmillan, 1948.

Nocks, Lisa. "The Origins and Development of the Magic Lantern." In *Science and Its Times: Understanding the Social Significance of Scientific Discovery*. Vol. 3, edited by Neil Schlager, 429-432. Detroit: Gale Group, 2000.

Núñez, Rafael E. "Conceptual Metaphor and the Embodied Mind: What Makes Mathematics Possible?" In *Metaphor and Analogy in the Sciences*, edited by Fernand Hallyn, 125-145. Dordrecht: Kluwer Academic Publishers, 2000.

Ovid. *Metamorphoses*. Translated by Horace Gregory. New York: Mentor, 1960.

Panofsky, Erwin. *Galileo as a Critic of the Arts*. The Hague: M. Nijhoff, 1954.

Paré, Ambroise. *On Monsters and Marvels*. Translated by Janis L. Pallister. Chicago: University of Chicago Press, 1983.

Paxson, James J. "Kepler's Allegory of Containment, the Making of Modern Astronomy, and the Semiotics of Mathematical Thought." *Intertexts* 3:2 (Fall 1999): 105-123.

———. *The Poetics of Personification*. Cambridge: Cambridge University Press, 1994.

———. "Revisiting the Deconstruction of Narratology: Master Tropes of Narrative Embedding and Symmetry." *Style* 35:1 (Spring 2001): 126-150.

Peck, Russell A. "Number as Cosmic Language." In *Essays in the Numerical Criticism of Medieval Literature*, edited by Caroline D. Eckhardt, 15-64. Lewisburg: Bucknell University Press, 1980.

Plato. *Plato's Cosmology*. Translated by F.M. Cornford. New York: Humanities Press, 1952.

[Plotinus.] Plotinos. *Complete Works*. Translated and edited by Kenneth Sylvan Guthrie. London, 1918.

Popper, Karl. *Conjectures and Refutations: The Growth of Scientific Knowledge*. New York: Harper & Row, 1963.

Price, S.R.F. "The Future of Dreams: From Freud to Artemidorus." *Past and Present* 113 (1986): 3-37.

Prickett, Stephen. *Narrative, Religion and Science: Fundamentalism Versus Irony, 1700-1999*. Cambridge: Cambridge University Press, 2002.

Ptolemy. *Ptolemy's Almagest*. Translated and edited by G.J. Toomer. New York: Springer Verlag, 1984.

Quilligan, Maureen. *The Language of Allegory: Defining the Genre*. Ithaca: Cornell University Press, 1979.

Randall, John Herman. "The Development of Scientific Method in the School of Padua." *Journal of the History of Ideas* 1 (1940): 177-206.

Reisenhuber, Klaus. "Arithmetic and the Metaphysics of Unity in Thierry of Chartres: On the Philosophy of Nature and Theology in the Twelfth Century." In *Nature in Medieval Thought: Some Approaches East and West*, edited by Chumaru Koyama, 43-73. Brill: Leiden, 2000.

Reiss, Timothy J. *The Discourse of Modernism*. Ithaca: Cornell University Press, 1982.

Resnik, Salomon. *La mise en scène du rêve*. Translated by Gabriella Mazzini. Paris: Payot, 1984.

Ricoeur, Paul. "Energetics and Hermeneutics in *The Interpretation of Dreams*." In *Modern Critical Interpretations: Sigmund Freud's The Interpretation of Dreams*, edited by Harold Bloom, 67-76. New York: Chelsea House, 1987.

Roberts, W. Rhys. *Longinus On the Sublime*. Cambridge: Cambridge University Press, 1907.

Roos, Anna Marie E. *Luminaries in the Natural World: The Sun and the Moon in England, 1400-1720*. New York: Peter Lang, 2001.

Rosen, Edward. *Three Imperial Mathematicians: Kepler Trapped Between Tycho Brahe and Ursus*. New York: Abaris, 1986.

Ryding, William W. *Structure in Medieval Narrative.* The Hague: Mouton, 1971.

Santillana, Giorgio de. *The Age of Adventure: The Renaissance Philosophers.* New York: Mentor, 1956.

Sautman, Francesca Canadé, Diana Conchado, and Giuseppe Carlo Di Scipio. "Texts and Shadows: Traces, Narratives, and Folklore." In *Telling Tales: Medieval Narratives and the Folk Tradition,* edited by Francesca Canadé Sautman, Diana Conchado, and Giuseppe Carlo Di Scipio, 1-20. New York: St. Martin's Press, 1998.

Schaeffer, Jean-Marie. *Qu'est-ce Qu'un Genre Littéraire?* Paris: Éditions du Seuil, 1989.

Schirrmeister, Albert. "Traum und Wissen in der Frühen Neuzeit." *Zeitsprünge* 5 (2001): 297-310.

Schmidt, Peter. *Het Lam Gods, Gent.* Gent: Ludion, 2001.

Schmitt, Charles B. "Philosophy and Science in Sixteenth-Century Universities: Some Preliminary Comments." In *Studies in Renaissance Philosophy and Science,* edited by Charles B. Schmitt, 485-530. London: Variorum Reprints, 1981.

Schmitt, Jean-Claude. *Ghosts in the Middle Ages: The Living and the Dead in Medieval Society.* Translated by Teresa Lavender Fagan. Chicago: University of Chicago Press, 1998.

Scholes, Robert and Robert Kellogg. *The Nature of Narrative.* New York: Oxford University Press, 1966.

Schrader, Ludwig. "Joachim Du Bellays 'Songe' (1558)—eine Vision des Untergangs." In *Traum und Träumen: Inhalt, Darstellung, Funktionen einer Lebenserfahrung in Mittelalter und Renaissance,* edited by Rudolf Hiestand. Düsseldorf: Droste Verlag, 1994.

Schrijvers, Piet. "Longinus and Quintilianus: The Classical Sublime." *Shadows of the Sublime: History of a Concept, the Third Gent Conference on Literary Theory.* Universiteit Gent, Gent. 12 Oct. 2002.

Serres, Michel. *Angels: A Modern Myth.* Translated by Francis Cowper. Paris: Flammarion, 1995.

Shea, William. "Looking at the Moon as another Earth. Terrestrial Analogies and Seventeenth-Century Telescopes." In *Metaphor and Analogy in the Sciences,* edited by Fernand Hallyn, 83-103. Dordrecht: Kluwer Academic Publishers, 2000.

Shipley, Joseph T. *Dictionary of Word Origins.* New York: Philosophical Library, 1945.

Simon, Gérard. "Analogies and Metaphors in Kepler." In *Metaphor and Analogy in the Sciences,* edited by Fernand Hallyn, 71-82. Dordrecht: Kluwer Academic Publishers, 2000.

Sophocles. *The Oedipus Cycle.* Translated by Dudley Fitts and Robert Fitzgerald. New York: Harvest, 1976.

Spearing, A.C. *Medieval-Dream Poetry.* Cambridge: Cambridge University Press, 1976.

Spenser, Edmund. *The Faerie Queene.* Edited by Thomas P. Roche, Jr. London: Penguin, 1987.

Steadman, John M. *Nature Into Myth: Medieval and Renaissance Moral Symbols.* Pittsburgh: Duquesne University Press, 1979.

Steel, Carlos. "Nature as Object of Science: On the Medieval Contribution to a Science of Nature." In *Nature in Medieval Thought: Some Approaches East and West*, edited by Chumaru Koyama, 125-152. Brill: Leiden, 2000.

Stephenson, Bruce. *Kepler's Physical Astronomy.* Princeton: Princeton University Press, 1994.

Stevens, Wesley M. "The Figure of the Earth in Isidore's *De natura rerum.*" *ISIS* 71 (1980): 268-277. Rpt. in *Cycles of Time and Scientific Learning in Medieval Europe*, edited by Wesley M. Stevens. Aldershot: Ashgate, 1995.

Stock, Brian. *Myth and Science in the Twelfth Century: A Study of Bernard Silvester.* Princeton: Princeton University Press, 1972.

Stock, R.D. *The Holy and the Daemonic from Sir Thomas Browne to William Blake.* Princeton: Princeton University Press, 1982.

Swinford, Dean. "Girard Desargues and Projective Geometry." In *Science and its Times: Understanding the Social Significance of Scientific Discovery. Vol 3: 2450-1699*, edited by Neil Schlager, 251-253. Detroit: Gale Group, 2000.

Szondi, Peter. *Introduction to Literary Hermeneutics.* Translated by Martha Woodmansee. ambridge: Cambridge University Press, 1995.

Teskey, Gordon. *Allegory and Violence.* Ithaca: Cornell University Press, 1996.

Thorndike, Lynn. *A History of Magic and Experimental Science: During the First Thirteen Centuries of Our Era.* Vol. II. New York: Columbia University Press, 1957.

———. *University Records and Life in the Middle Ages.* New York: Columbia University Press, 1949.

Todorov, Tzvetan. *Introduction á la Littérature Fantastique.* Paris: Éditions du Seuil, 1970.

———. *The Poetics of Prose.* Translated by Richard Howard. Ithaca: Cornell University Press, 1977.

Trilling, Lionel. *The Liberal Imagination.* New York: Doubleday, 1953.

Tulloch, David. "Isaac Newton's *Principia Mathematica* Greatly Influences the Scientific World and the Society Beyond It." In *Science and its Times: Understanding the Social Significance of Scientific Discovery. Vol 3: 2450-1699*, edited by Neil Schlager, 326-329. Detroit: Gale Group, 2000.

Tuve, Rosemond. *Allegorical Imagery: Some Mediaeval Books and Their Posterity.* Princeton: Princeton University Press, 1966.

Vance, Eugene. *Mervelous Signals: Poetics and Sign Theory in the Middle Ages.* Lincoln: University of Nebraska Press, 1986.

Van Lieshout, R.G.A. *Greeks On Dreams.* Utrecht: HES Publishers, 1980.

Vassilopoulou, Panayiota. "Creation or Metamorphosis? Plotinus on the Genesis of the World." In *Neoplatonism and Western Aesthetics*, edited by Aphrodite Alexandrakis, 207-228. Albany: SUNY Press, 2001.

Viëtor, Karl. "L'histoire des genres littéraires." Translated by Jean-Pierre Morel. *Théorie des Genres*, edited by Gérard Genette and Tzvetan Todorov. Paris: Éditions du Seuil, 1986.

Virgil. *The Aeneid.* Trans. Allen Mandelbaum. New York: Bantam, 1981.

Voelkel, James R. *Johannes Kepler and the New Astronomy.* New York: Oxford University Press, 1999.

Vossler, Karl. *Mediaeval Culture: An Introduction to Dante and His Times.* Vol. 1. Translated by William Cranston Lawton. New York: Frederick Ungar Publishing Co., 1966.

———. *Mediaeval Culture: An Introduction to Dante and His Times.* Vol. 2. Translated by William Cranston Lawton. New York: Frederick Ungar Publishing Co., 1966.

Waddell, Helen. *The Wandering Scholars: The Life and Art of the Lyric Poets of the Latin Middle Ages.* New York: Doubleday, 1955.

Watson, Stephanie. "Camera Obscura: Ancestor of Modern Photography." In *Science and its Times: Understanding the Social Significance of Scientific Discovery. Vol 3: 2450-1699*, edited by Neil Schlager, 423-426. Detroit: Gale Group, 2000.

Weidhorn, M. *Dreams in Seventeenth-Century Literature.* The Hague: Mouton, 1970.

Wellek, René and Austin Warren. *Theory of Literature.* 3rd ed. New York: Harvest, 1956.

Westfall, Richard S. *The Construction of Modern Science: Mechanisms and Mechanics.* Cambridge: Cambridge University Press, 1977.

Westman, Robert S. "Michael Maestlin's Adoption of the Copernican Theory." *Studia Copernicana* XIV (1975): 53-63.

Wetherbee, Winthrop. "Philosophy, Cosmology, and the Twelfth-Century Renaissance." In *A History of Twelfth-Century Western Philosophy*, edited by Peter Dronke, 21-53. Cambridge: Cambridge University Press, 1988. 21-53.

Whitaker, Ewen A. "Selenography in the Seventeenth Century." In *Planetary Astronomy from the Renaissance to the Rise of Astrophysics. Part A: Tycho Brahe to Newton*, edited by R. Taton and C. Wilson. Cambridge: Cambridge University Press, 1989.

Whitman, Jon. *Allegory: The Dynamics of an Ancient and Medieval Technique.* Oxford: Clarendon Press, 1987.

———. *Interpretation and Allegory: Antiquity to the Modern Period.* Leiden: Brill, 2000.

Wieruszowski, Helene. *The Medieval University.* Princeton: Anvil, 1966.

Wilson, David L. and Zack Bowen. *Science and Literature: Bridging the Two Cultures.* Gainesville: University Press of Florida, 2001.

Wittkower, Rudolf. *Architecture in the Age of Humanism.* 1949. Reprint, London: Academy Editions, 1988.

Index

E

F

O

P

www.ingramcontent.com/pod-product-compliance
Ingram Content Group UK Ltd.
Pitfield, Milton Keynes, MK11 3LW, UK
UKHW022000180926
475701UK00003B/118
9780415886079